DC Electric Machines, Electromechanical Energy Conversion Principles, and Magnetic Circuit Analysis

Mehdi Rahmani-Andebili

DC Electric Machines, Electromechanical Energy Conversion Principles, and Magnetic Circuit Analysis

Practice Problems, Methods, and Solutions

 Springer

Mehdi Rahmani-Andebili
Department of Engineering and Physics
University of Central Oklahoma
Edmond, OK, USA

ISBN 978-3-031-08865-0   ISBN 978-3-031-08863-6   (eBook)
https://doi.org/10.1007/978-3-031-08863-6

© The Editor(s) (if applicable) and The Author(s), under exclusive license to Springer Nature Switzerland AG 2022
This work is subject to copyright. All rights are solely and exclusively licensed by the Publisher, whether the whole or
part of the material is concerned, specifically the rights of translation, reprinting, reuse of illustrations, recitation,
broadcasting, reproduction on microfilms or in any other physical way, and transmission or information storage and
retrieval, electronic adaptation, computer software, or by similar or dissimilar methodology now known or hereafter
developed.
The use of general descriptive names, registered names, trademarks, service marks, etc. in this publication does not
imply, even in the absence of a specific statement, that such names are exempt from the relevant protective laws and
regulations and therefore free for general use.
The publisher, the authors, and the editors are safe to assume that the advice and information in this book are believed to
be true and accurate at the date of publication. Neither the publisher nor the authors or the editors give a warranty,
expressed or implied, with respect to the material contained herein or for any errors or omissions that may have been
made. The publisher remains neutral with regard to jurisdictional claims in published maps and institutional affiliations.

This Springer imprint is published by the registered company Springer Nature Switzerland AG
The registered company address is: Gewerbestrasse 11, 6330 Cham, Switzerland

Preface

The courses of DC electric machines and electromechanical energy conversion principles are two of the fundamental courses of electric power engineering major that are taught for junior students. The subjects include magnetic circuits analysis, inductance of windings and induced voltage in windings, energy loss in magnetic circuits, electromechanical energy conversion, separately excited DC electric machines, shunt DC electric machines, series DC electric machines, compound DC electric machines, and power loss and efficiency of DC electric machines.

Like the author's previously published textbooks, this textbook includes very detailed and multiple methods of problem solutions. It can be used as a practicing textbook by students and as a supplementary teaching source by instructors.

To help students study the textbook in the most efficient way, the exercises have been categorized in nine different levels. In this regard, for each problem of the textbook, a difficulty level (easy, normal, or hard) and a calculation amount (small, normal, or large) have been assigned. Moreover, in each chapter, problems have been ordered from the easiest problem with the smallest calculations to the most difficult problems with the largest calculations. Therefore, students are suggested to start studying the textbook from the easiest problems and continue practicing until they reach the normal and then the hardest ones. On the other hand, this classification can help instructors choose their desirable problems to conduct a quiz or a test. Moreover, the classification of computation amount can help students manage their time during future exams and instructors give the appropriate problems based on the exam duration.

Since the problems have very detailed solutions and some of them include multiple methods of solution, the textbook can be useful for the under-prepared students. In addition, the textbook is beneficial for knowledgeable students because it includes advanced exercises.

In the preparation of problem solutions, an attempt has been made to use typical methods of electrical circuit analysis to present the textbook as an instructor-recommended one. In other words, the heuristic methods of problem solution have never been used as the first method of problem solution. By considering this key point, the textbook will be in the direction of instructors' lectures, and the instructors will not see any untaught problem solutions in their students' answer sheets.

The Iranian University Entrance Exams for the master's and PhD degrees of electrical engineering major is the main reference of the textbook; however, all the problem solutions have been provided by me. The Iranian University Entrance Exam is one of the most competitive university entrance exams in the world that allows only 10% of the applicants to get into prestigious and tuition-free Iranian universities.

The author has already published the below books and textbooks with Springer.

Textbooks

Differential Equations: Practice Problems, Methods, and Solutions, 2022.
Feedback Control Systems Analysis and Design: Practice Problems, Methods, and Solutions, 2022.
Power System Analysis: Practice Problems, Methods, and Solutions, 2022.
Advanced Electrical Circuit Analysis: Practice Problems, Methods, and Solutions, 2022.
AC Electrical Circuit Analysis: Practice Problems, Methods, and Solutions, 2021.
Calculus: Practice Problems, Methods, and Solutions, 2021.
Precalculus: Practice Problems, Methods, and Solutions, 2021.
DC Electrical Circuit Analysis: Practice Problems, Methods, and Solutions, 2020.

Books

Applications of Artificial Intelligence in Planning and Operation of Smart Grid, 2022.
Design, Control, and Operation of Microgrids in Smart Grids, 2021.
Applications of Fuzzy Logic in Planning and Operation of Smart Grids, 2021.
Operation of Smart Homes, 2021.
Planning and Operation of Plug-in Electric Vehicles: Technical, Geographical, and Social Aspects, 2019.

Edmond, OK, USA   Mehdi Rahmani-Andebili

Contents

About the Author

Mehdi Rahmani-Andebili is an assistant professor in the Department of Engineering and Physics at the University of Central Oklahoma, OK, USA. Before that, he was also an assistant professor in the Electrical Engineering Department at Montana Technological University, MT, USA, and the Engineering Technology Department at State University of New York, Buffalo State, NY, USA, during 2019–2022. He received his first MSc and PhD degrees in electrical engineering (power system) from Tarbiat Modares University and Clemson University in 2011 and 2016, respectively, and his second MSc degree in physics and astronomy from the University of Alabama in Huntsville in 2019. Moreover, he was a postdoctoral fellow at Sharif University of Technology during 2016–2017. As a professor, he has taught many courses and labs, including power system analysis, DC and AC electric machines, feedback control systems analysis and design, renewable distributed generation and storage, industrial electronics, analog electronics, electrical circuits and devices, AC electrical circuits analysis, DC electrical circuits analysis, essentials of electrical engineering technology, and algebra- and calculus-based physics. Dr. Rahmani-Andebili has more than 200 single-author and first-author publications, including journal papers, conference papers, textbooks, books, and book chapters. He is an IEEE Senior Member and the permanent reviewer of many credible journals. His research areas include smart grid, power system operation and planning, integration of renewables and energy storages into power system, energy scheduling and demand-side management, plug-in electric vehicles, distributed generation, and advanced optimization techniques in power system studies.

Problems: Magnetic Circuits Analysis

Abstract

In this chapter, Kirchhoff's Magnetomotive Force Law (KML) and Kirchhoff's Flux Law (KFL) are applied on the equivalent electrical circuits of the magnetic circuits to analyze their basic and advanced problems. In this chapter, the problems are categorized in different levels based on their difficulty levels (easy, normal, and hard) and calculation amounts (small, normal, and large). Additionally, the problems are ordered from the easiest problem with the smallest computations to the most difficult problems with the largest calculations.

1.1. Figure 1.1.a shows a magnetic core that its magnetization curve is illustrated in Fig. 1.1.a. The cross-sectional area and the average perimeter of the core are about 20 cm^2 and 50 cm, respectively. Determine the number of turns of coil C to create the magnetic flux of 3.5 mWb in the core. Herein, assume that $\mu_0 \approx 10^{-6}$ H/m.

Difficulty level ○ Easy ● Normal ○ Hard
Calculation amount ○ Small ● Normal ○ Large

1) 275
2) 150
3) 350
4) 125

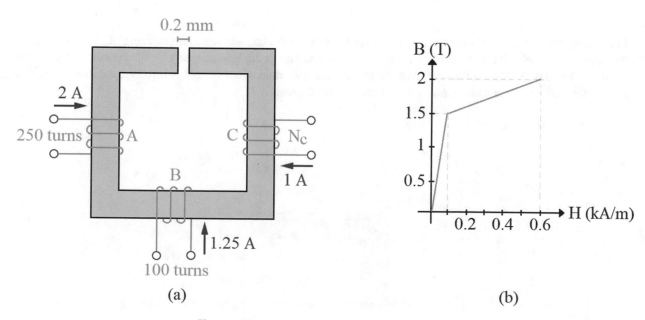

Fig. 1.1 The magnetic circuit and graph of problem 1.1

© The Author(s), under exclusive license to Springer Nature Switzerland AG 2022
M. Rahmani-Andebili, *DC Electric Machines, Electromechanical Energy Conversion Principles, and Magnetic Circuit Analysis*,
https://doi.org/10.1007/978-3-031-08863-6_1

1.2. In the magnetic circuit shown in Fig. 1.2, the relative permeability of the core is about 20,000. Determine the ampere-turn of the coil to create the magnetic flux density of 1 T in the core. Herein, assume that the fringing factor is about 1.05.

Difficulty level   ○ Easy   ● Normal   ○ Hard
Calculation amount   ○ Small   ● Normal   ○ Large

1) 2796 *AT*
2) 3876 *AT*
3) 3020 *AT*
4) 7608 *AT*

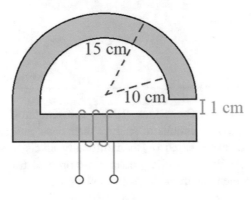

Fig. 1.2 The magnetic circuit of problem 1.2

1.3. In a circular object made from a magnetic material with the relative permeability of 1000, an air gap of 2 mm is created. To keep the primary magnetic flux unchanged, what must be the ratio of the new magnetomotive force (mmf) to the primary value if the average perimeter of the object is 1 m?

Difficulty level   ○ Easy   ● Normal   ○ Hard
Calculation amount   ○ Small   ● Normal   ○ Large

1) 3
2) 2
3) 1
4) 1.5

1.4. In the magnetic circuit shown in Fig. 1.3, the number of turns of the coil and the average length of the core are 500 and 360 mm, respectively. If the required magnetic flux density and magnetic field intensity for the operation of the relay are about 0.8 T and 510 AT/m in the core, respectively, calculate the ratio of the current, when there are two air gaps of 1.5 mm, to the current, when there is no air gap in the magnetic circuit.

Difficulty level   ○ Easy   ● Normal   ○ Hard
Calculation amount   ○ Small   ● Normal   ○ Large

1) 8
2) 13
3) 11
4) 15

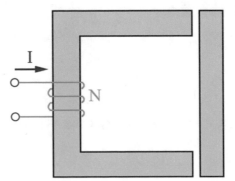

Fig. 1.3 The magnetic circuit of problem 1.4

1.5. In the magnetic circuit illustrated in Fig. 1.4, $N_1 = 300$, $N_2 = 500$, and $N_3 = 350$. In addition, $l_{g1} = \sqrt{2}l_{g2}$, $A_1 = \frac{1}{\sqrt{2}}A_2$, and $I_1 = I_2 = 1\,A$. For what value of I_3, the magnetic flux flowing through the middle column is zero? Herein, assume that the permeability of the core is infinite.

Difficulty level   ○ Easy   ○ Normal   ● Hard
Calculation amount   ○ Small   ○ Normal   ● Large

1) $-\frac{2}{3}$

2) $-\frac{3}{2}$

3) $\frac{2}{3}$

4) $\frac{3}{2}$

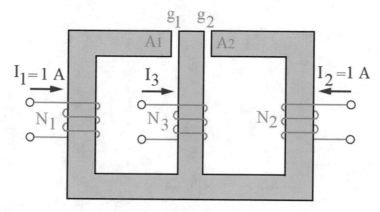

Fig. 1.4 The magnetic circuit of problem 1.5

Abstract

In this chapter, the problems of the first chapter are fully solved, in detail, step-by-step, and with different methods.

2.1. Based on the information given in the problem, we have:

$$A = 20 \; cm^2 = 20 \times 10^{-4} \; m^2 \tag{1}$$

$$l_c = 0.5 \; m \tag{2}$$

$$l_g = 0.2 \times 10^{-3} \; m \tag{3}$$

$$\varphi_c = 3.5 \times 10^{-3} \; Wb \tag{4}$$

$$N_A = 250, \quad N_B = 100 \tag{5}$$

$$I_A = 2 \; A, \quad I_B = 1.25 \; A, \quad I_C = 1 \; A \tag{6}$$

$$\mu_0 \approx 10^{-6} \frac{H}{m} \tag{7}$$

As we know, by writing Kirchhoff's Magnetomotive Force Law (KML) for a magnetic circuit and considering the direction of magnetic flux of coils, we have:

$$\sum_{i=1}^{m} F_{m,i} = \sum_{j=1}^{n} V_{m,j} \tag{8}$$

where:

$$\sum_{i=1}^{m} F_{m,i} = \sum_{i=1}^{m} N_i I_i \tag{9}$$

$$\sum_{j=1}^{n} V_{m,j} = \sum_{j=1}^{n} R_j \varphi_j = \sum_{j=1}^{n} H_j l_j \tag{10}$$

Equations (8)–(10) state that the sum of magnetomotive forces (mmf) in a loop ($\sum_{i=1}^{m} F_{m,i}$), achieved from the ampere-turn excitation of the coils, is equal to sum of mmf drops ($\sum_{j=1}^{n} V_{m,j}$), achieved from the product of flux and reluctances or the product of magnetic field intensity and length of magnetic components, across the rest of the loop.

© The Author(s), under exclusive license to Springer Nature Switzerland AG 2022
M. Rahmani-Andebili, *DC Electric Machines, Electromechanical Energy Conversion Principles, and Magnetic Circuit Analysis*,
https://doi.org/10.1007/978-3-031-08863-6_2

Now, by applying (8)–(10) on the magnetic circuit of Fig. 2.1, we can write:

$$\Rightarrow (N_A I_A - N_B I_B + N_C I_C) = H_c l_c + H_{ag} l_{ag} \tag{11}$$

As can be noticed from Fig. 2.1, the direction of magnetic flux of coils A and C is clockwise but the direction of magnetic flux of coil B is counterclockwise.

Since the cross-sectional area of the magnetic circuit in the core is the same as in the air gap, the magnetic field density in them will be equal, as is presented in the following:

$$\Rightarrow B_c = B_{ag} = \frac{\varphi}{A} = \frac{3.5 \times 10^{-3}}{20 \times 10^{-4}} = 1.75\ T \tag{12}$$

From the second segment of the magnetization curve of the core, we have the relation below:

$$B_c = H_c + 1.4 \tag{13}$$

In (13), B_c and H_c are in T and kA/m, respectively.

Solving (12) and (13):

$$\Rightarrow H_c = 0.35 \frac{kA}{m} = 350 \frac{A}{m} \tag{14}$$

Moreover, the magnetic field intensity in the air gap can be calculated by using (7) and (12) as follows:

$$H_{ag} = \frac{B_{ag}}{\mu_0} = \frac{1.75}{10^{-6}} = 1.75 \times 10^6 \frac{A}{m} \tag{15}$$

Solving (11) by considering all the quantities given:

$$(250 \times 2) - (100 \times 1.25) + (N_c \times 1) = (350 \times 0.5\ ) + (1.75 \times 10^6 \times 0.2 \times 10^{-3}) \tag{14}$$

$$\Rightarrow N_C = 150$$

Choice (2) is the answer.

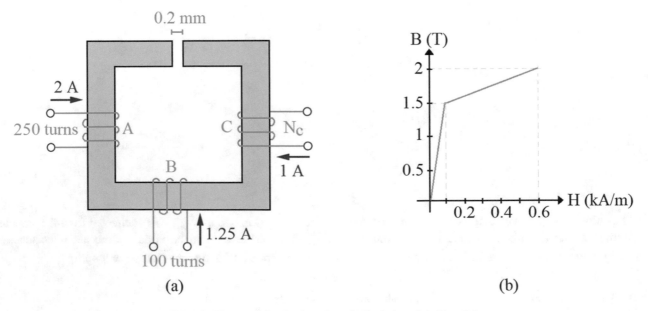

(a)   (b)

Fig. 2.1 The magnetic circuit and graph of solution of problem 2.1

2.2. Based on the information given in the problem, we have:

$$\mu_r = 20000 \tag{1}$$

$$B_c = 1\,T \tag{2}$$

$$\text{fringing factor} = 1.05 \tag{3}$$

The problem can be solved as follows:

$$B_{ag} = \frac{B_c}{\text{fringing factor}} = \frac{1}{1.05} = 0.952\,T \tag{4}$$

By writing Kirchhoff's Magnetomotive Force Law (KML) for a magnetic circuit, we have:

$$\sum_{i=1}^{m} F_{m,i} = \sum_{j=1}^{n} V_{m,j} \Rightarrow \sum_{i=1}^{m} N_i I_i = \sum_{j=1}^{n} H_j l_j \tag{5}$$

Equation (5) states that the sum of magnetomotive forces (mmf) in a loop ($\sum_{i=1}^{m} F_{m,i}$), achieved from the ampere-turn excitation of the coils, is equal to the sum of mmf drops ($\sum_{j=1}^{n} V_{m,j}$), achieved from the product of magnetic field intensity and length of magnetic components, across the rest of the loop.

Now, by applying (5) on the magnetic circuit, we can write:

$$\Rightarrow NI = H_c l_c + H_{ag} l_{ag} \xrightarrow{H = \frac{B}{\mu}} NI = \frac{B_c}{\mu_0 \mu_r} l_c + \frac{B_{ag}}{\mu_0} l_{ag} \tag{6}$$

The average length of the core can be calculated as follows:

$$l_c = \frac{2\pi r_{av}}{2} + 2r_{av} + 1 + 2 \times \frac{5}{2} = \pi \times \left(\frac{10+15}{2}\right) + 2 \times \left(\frac{10+15}{2}\right) + 1 + 5 = 70.27\,cm \tag{7}$$

Solving (6) by considering all the quantities given:

$$NI = \left(\frac{1}{4\pi \times 10^{-7} \times 20000} \times 0.7027\right) + \left(\frac{0.952}{4\pi \times 10^{-7}} \times 0.01\right) = 27.97 + 7579.62 \tag{8}$$

$$\Rightarrow NI \approx 7608\,AT$$

Choice (4) is the answer.

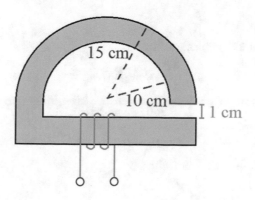

Fig. 2.2 The magnetic circuit of solution of problem 2.2

2.3. Based on the information given in the problem, we have:

$$\mu_r = 1000 \tag{1}$$

$$l_{ag} = 2\ mm = 0.002\ m \tag{2}$$

$$\varphi_1 = \varphi_2 \tag{3}$$

$$l_c = 1\ m \tag{4}$$

In the first condition (the core without any air gap), the magnetic flux can be calculated by writing Kirchhoff's Magnetomotive Force Law (KML) for the magnetic circuit, as follows:

$$\sum_{i=1}^{m} F_{m,i} = \sum_{j=1}^{n} V_{m,j} \Rightarrow \sum_{i=1}^{m} N_i I_i = \sum_{j=1}^{n} R_j \varphi_j \tag{5}$$

Equation (5) states that sum of magnetomotive forces (mmf) in a loop ($\sum_{i=1}^{m} F_{m,i}$), achieved from the ampere-turn excitation of the coils, is equal to the sum of mmf drops ($\sum_{j=1}^{n} V_{m,j}$), achieved from the product of magnetic field intensity and length of magnetic components, across the rest of the loop:

$$\varphi_1 = \frac{NI_1}{R_1} = \frac{NI_1}{R_c} = \frac{NI_1}{\frac{l_c}{\mu_0 \mu_r A}} \tag{6}$$

$$\Rightarrow \varphi_1 = \frac{NI_1}{\frac{1}{1000\mu_0 A}} = 1000\mu_0 A N I_1 \tag{7}$$

In the second condition (the core with the air gap), the magnetic flux can be calculated as follows:

$$\varphi_2 = \frac{NI_2}{R_2} = \frac{NI_2}{R_c + R_{ag}} = \frac{NI_2}{\frac{l_c}{\mu_0 \mu_r A} + \frac{l_g}{\mu_0 A}} \tag{8}$$

$$\Rightarrow \varphi_2 = \frac{NI_2}{\frac{1}{1000\mu_0 A} + \frac{2 \times 10^{-3}}{\mu_0 A}} = \frac{NI_2}{\frac{3}{1000\mu_0 A}} = \frac{1000}{3}\mu_0 A N I_2 \tag{9}$$

Solving (3), (7), and (9):

$$1000\mu_0 A N I_1 = \frac{1000}{3}\mu_0 A N I_2 \Rightarrow \frac{F_{m,2}}{F_{m,1}} = \frac{NI_2}{NI_1} = 3$$

Choice (1) is the answer.

2.4. Based on the information given in the problem, we have:

$$N = 500 \tag{1}$$

$$l_c = 360\ mm = 0.36\ m \tag{2}$$

$$B_c = 0.8\ T \tag{3}$$

$$H_c = 510 \, \frac{AT}{m} \tag{4}$$

$$l_{ag} = 1.5 \, mm = 1.5 \times 10^{-3} \, m \tag{5}$$

By writing Kirchhoff's Magnetomotive Force Law (KML) for a magnetic circuit, we have:

$$\sum_{i=1}^{m} F_{m,i} = \sum_{j=1}^{n} V_{m,j} \Rightarrow \sum_{i=1}^{m} N_i I_i = \sum_{j=1}^{n} H_j l_j \tag{6}$$

Equation (6) states that the sum of magnetomotive forces (mmf) in a loop ($\sum_{i=1}^{m} F_{m,i}$), achieved from the ampere-turn excitation of the coils, is equal to the sum of mmf drops ($\sum_{j=1}^{n} V_{m,j}$), achieved from the product of magnetic field intensity and length of magnetic components, across the rest of the loop.

Now, by applying (6) on the magnetic circuit of Fig. 2.3, when there are two air gaps, we have:

$$NI_1 = H_c l_c + H_{ag} l_{ag} \xrightarrow{H_{ag} = \frac{B_{ag}}{\mu_0}} NI_1 = H_c l_c + \frac{B_{ag}}{\mu_0} l_{ag} \xrightarrow{B_c = B_{ag}} NI_1 = H_c l_c + \frac{B_c}{\mu_0} l_{ag} \tag{7}$$

$$\Rightarrow 500 I_1 = 510 \times 0.36 + \frac{0.8}{4\pi \times 10^{-7}} \times 2 \times 1.5 \times 10^{-3} \Rightarrow I_1 = 4.2 \, A \tag{8}$$

When there is no air gap in the system:

$$NI_2 = H_c l_c$$

$$\Rightarrow 500 I_2 = 510 \times 0.36 \Rightarrow I_2 = 0.3672 A \tag{9}$$

$$\frac{I_1}{I_2} = \frac{4.2}{0.3672} \approx 11$$

Choice (3) is the answer.

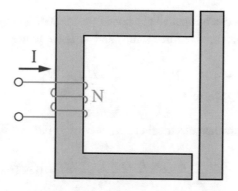

Fig. 2.3 The magnetic circuit of solution of problem 2.4

2.5. Based on the information given in the problem, we have:

$$N_1 = 300, \quad N_2 = 500, \quad N_3 = 350 \tag{1}$$

$$l_{g1} = \sqrt{2} l_{g2} \tag{2}$$

$$A_1 = \frac{1}{\sqrt{2}} A_2 \tag{3}$$

$$I_1 = I_2 = 1 \, A \tag{4}$$

$$\varphi_3 = 0 \tag{5}$$

$$\mu_c = \infty \tag{6}$$

The reluctance of the first air gap can be calculated as follows:

$$R_{g1} = \frac{l_{g1}}{\mu_0 A_1} \tag{7}$$

Likewise, the reluctance of the second air gap can be calculated as follows:

$$R_{g2} = \frac{l_{g2}}{\mu_0 A_2} \tag{8}$$

Solving (2), (3), and (7):

$$R_{g1} = \frac{\sqrt{2} l_{g2}}{\mu_0 \frac{1}{\sqrt{2}} A_2} = 2 \frac{l_{g2}}{\mu_0 A_2} \tag{9}$$

Solving (8) and (9):

$$R_{g1} = 2 R_{g2} \tag{9}$$

As we know, by writing Kirchhoff's Flux Law (KFL) at a node and considering the direction of magnetic flux of each branch, we have:

$$\sum_{k=1}^{q} \varphi_k = 0 \tag{10}$$

In addition, because of $\varphi = \frac{F}{R}$, KFL can be applied at a node (e.g., at node x connected to some other nodes), as follows, where $F = NI$ represents the magnetomotive force (mmf) and R_k is the reluctance of k'th branch:

$$\sum_{k=1}^{q} \frac{F_x - F_k}{R_k} = 0 \Rightarrow \sum_{k=1}^{q} \frac{N_x I_x - N_k I_k}{R_k} = 0 \tag{11}$$

Applying (11) at node 1 of the equivalent electrical circuit, shown in Fig. 2.4.b, and considering all the quantities:

$$\frac{N_3 I_3 - N_1 I_1}{2 R_{g2}} + \frac{N_3 I_3 - (-N_2 I_2)}{R_{g2}} + 0 = 0 \tag{12}$$

$$\Rightarrow \frac{350 I_3 - 300}{2 R_{g2}} + \frac{350 I_3 + 500}{R_{g2}} = 0 \Rightarrow 175 I_3 - 150 + 350 I_3 + 500 = 0 \Rightarrow 525 I_3 = -350$$

$$\Rightarrow I_3 = -\frac{2}{3} \, A$$

Choice (1) is the answer.

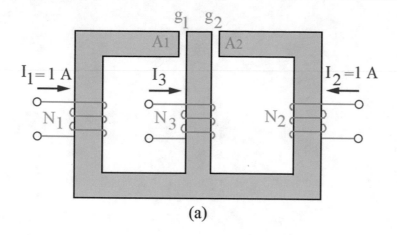

(a)

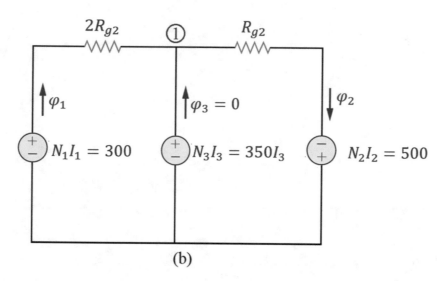

(b)

Fig. 2.4 The magnetic circuit and equivalent electrical circuit of solution of problem 2.5

Problems: Inductance of Windings and Induced Voltage in Windings

3

Abstract

In this chapter, the self-inductance and mutual inductance of the windings and coils are determined, and the voltage induced in them is calculated using Faraday's law. In this chapter, the problems are categorized in different levels based on their difficulty levels (easy, normal, and hard) and calculation amounts (small, normal, and large). Additionally, the problems are ordered from the easiest problem with the smallest computations to the most difficult problems with the largest calculations.

3.1. In the magnetic circuit of Fig. 3.1, calculate the inductance of the coil if the inner radius, the outer radius, and the height of the core are 80 mm, 100 mm, and 20 mm, respectively. Moreover, the number of turns of the coil and the relative permeability of the core are 200 and 900, respectively.

Difficulty level   ○ Easy   ● Normal   ○ Hard
Calculation amount   ● Small   ○ Normal   ○ Large

1) 0.032 *H*
2) 0.32 *H*
3) 0.23 *H*
4) 0.23 *mH*

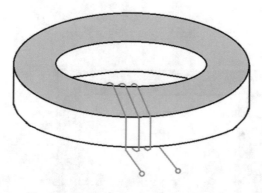

Fig. 3.1 The magnetic circuit of problem 3.1

3.2. In the magnetic circuit shown in Fig. 3.2, the core is ideal, and the resistance of the coils is zero. If $v(t) = 2V_m \sin \omega t$ and the cross-sectional area of the columns is the same, what is the relation between the effective value of the induced voltages (V_1 and V_2)?

Difficulty level   ○ Easy   ○ Normal   ● Hard
Calculation amount   ● Small   ○ Normal   ○ Large

© The Author(s), under exclusive license to Springer Nature Switzerland AG 2022
M. Rahmani-Andebili, *DC Electric Machines, Electromechanical Energy Conversion Principles, and Magnetic Circuit Analysis*,
https://doi.org/10.1007/978-3-031-08863-6_3

1) $\frac{V_1}{V_2} = \left(\frac{N_1}{N_2}\right)^2$

2) $\frac{V_1}{V_2} = \frac{N_1}{N_2}$

3) $\frac{V_1}{V_2} = \frac{N_2}{N_1}$

4) $\frac{V_1}{V_2} = \left(\frac{N_2}{N_1}\right)^2$

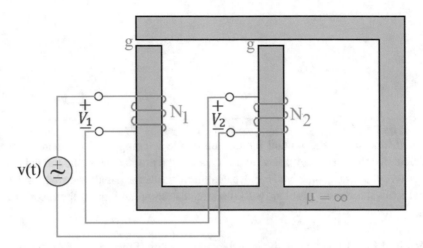

Fig. 3.2 The magnetic circuit of problem 3.2

3.3. In the electromagnetic relay shown in Fig. 3.3, the core is ideal. If the right-hand side part of the relay moves with the linear velocity of v, calculate the voltage induced in the open-circuit coil. Herein, consider that the velocity is slow enough so that the current of the first coil can be assumed constant.

Difficulty level   ○ Easy   ○ Normal   ● Hard
Calculation amount   ● Small   ○ Normal   ○ Large

1) $k\frac{v^2}{x^2}$

2) $k\frac{v}{x^2}$

3) $k\frac{x^2}{v}$

4) kvx

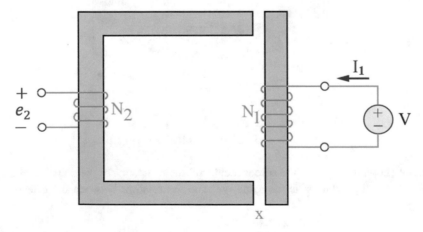

Fig. 3.3 The magnetic circuit of problem 3.3

3.4. In the magnetic circuit illustrated in Fig. 3.4, if the cross-sectional area of the core is the same everywhere and the permeability of the core is infinite, what is the value of $\frac{L_{22}}{L_{12}}$?

Difficulty level   ○ Easy   ○ Normal   ● Hard
Calculation amount   ○ Small   ● Normal   ○ Large

1) $\frac{N_2}{N_1}$

2) $\frac{N_1}{2N_2}$

3) $\frac{N_1}{N_2}$

4) $\frac{2N_2}{N_1}$

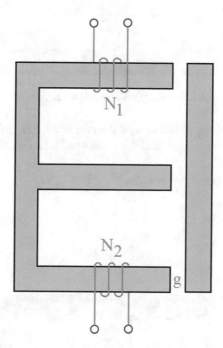

Fig. 3.4   The magnetic circuit of problem 3.4

3.5. In the magnetic circuit of Fig. 3.5, if the cross-sectional area of the columns is the same everywhere (A) and the permeability of the core is infinite, calculate the mutual inductance between the coils.

Difficulty level   ○ Easy   ○ Normal   ● Hard
Calculation amount   ○ Small   ● Normal   ○ Large

1) $\frac{\mu_0 A N_1 N_2}{2g}$

2) $\frac{\mu_0 A N_1 N_2}{g}$

3) $\frac{\mu_0 A N_1 N_2^2}{2g}$

4) $\frac{\mu_0 A N_1 N_2^2}{g}$

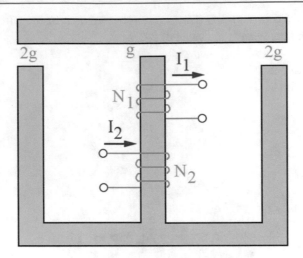

Fig. 3.5 The magnetic circuit of problem 3.5

3.6. In the magnetic circuit of Fig. 3.6, if the inductance of the coil is 10 mH, calculate the length of the air gap. Herein, assume that the reluctance of the core is zero and the cross-sectional area of the middle and the outer columns are 400 mm^2 and 300 mm^2, respectively.

Difficulty level   ○ Easy   ○ Normal   ● Hard
Calculation amount   ○ Small   ● Normal   ○ Large

1) $100\mu_0$
2) $600\mu_0$
3) $130\mu_0$
4) $240\mu_0$

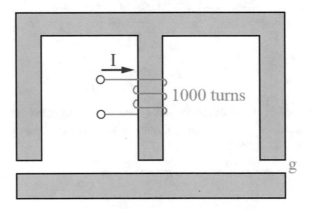

Fig. 3.6 The magnetic circuit of problem 3.6

3.7. In the magnetic circuit of Fig. 3.7, the cross-sectional area of the middle columns is two times of the one of the outer columns. What is the relation between the self-inductance (L_{aa}) and the mutual inductance (L_{ab}) in coils "a" and "b" if $N_a = N_b = N_c$ and the reluctance of the horizontal branches is zero?

Difficulty level   ○ Easy   ○ Normal   ● Hard
Calculation amount   ○ Small   ○ Normal   ● Large

1) $L_{aa} = 2L_{ab}$
2) $L_{aa} = 3L_{ab}$
3) $L_{aa} = 4L_{ab}$
4) $L_{aa} = 1.5L_{ab}$

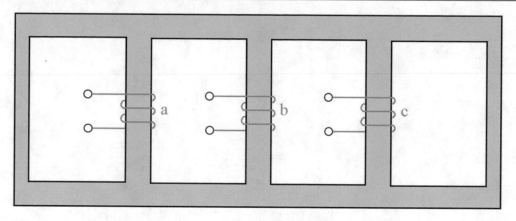

Fig. 3.7 The magnetic circuit of problem 3.7

Solutions of Problems: Inductance of Windings and Induced Voltage in Windings

Abstract

In this chapter, the problems of the third chapter are fully solved, in detail, step-by-step, and with different methods.

4.1. Based on the information given in the problem, we have:

$$r_{inner} = 80 \ mm \tag{1}$$

$$r_{outer} = 100 \ mm \tag{2}$$

$$h = 20 \ mm \tag{3}$$

$$N = 200 \tag{4}$$

$$\mu_r = 900 \tag{5}$$

The reluctance seen by the coil can be calculated as follows:

$$R = \frac{l_{av}}{\mu A} = \frac{2\pi r_{av}}{\mu_0 \mu_r A} = \frac{2\pi \times \frac{r_{outer} + r_{inner}}{2}}{\mu_0 \mu_r \times (r_{outer} - r_{inner}) \times h}$$

$$\Rightarrow R = \frac{2\pi \times \frac{100 + 80}{2} \times 10^{-3}}{4\pi \times 10^{-7} \times 900 \times (100 - 80) \times 10^{-3} \times 20 \times 10^{-3}} = 125 \times 10^4 \ \frac{A}{Wb}$$

The inductance of the coil can be calculated as follows:

$$L = \frac{N^2}{R} = \frac{200^2}{125 \times 10^4} \Rightarrow L = 0.032 \ H$$

Choice (1) is the answer.

© The Author(s), under exclusive license to Springer Nature Switzerland AG 2022
M. Rahmani-Andebili, *DC Electric Machines, Electromechanical Energy Conversion Principles, and Magnetic Circuit Analysis*,
https://doi.org/10.1007/978-3-031-08863-6_4

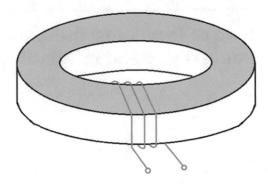

Fig. 4.1 The magnetic circuit of solution of problem 4.1

4.2. Based on the information given in the problem, we know that the resistance of the coils is zero, the cross-sectional area of the columns is the same, and the core is ideal. Therefore, only the air gaps have a reluctance. Moreover, we know that:

$$v(t) = 2V_m \sin \omega t \tag{1}$$

Figure 4.2.b shows the equivalent electrical circuit of the magnetic circuit. Since there is a short circuit branch parallel to both branches, the whole magnetic flux of each coil flows through the right-hand side branch. Therefore, there is no mutual inductance between the coils ($L_{12} = L_{21} = 0$) and the inductance of each coil is as follows:

$$L_{11} = \frac{N_1^2}{R_g}, \quad L_{22} = \frac{N_2^2}{R_g} \tag{2}$$

The equations below present the voltage of the coils:

$$\begin{cases} v_1(t) = L_{11}\frac{di}{dt} + \underbrace{L_{12}\frac{di}{dt}}_{0} = L_{11}\frac{di}{dt} \\[2em] v_2(t) = \underbrace{L_{21}\frac{di}{dt}}_{0} + L_{22}\frac{di}{dt} = L_{22}\frac{di}{dt} \end{cases} \tag{3}$$

$$\frac{v_1(t)}{v_2(t)} = \frac{L_{11}}{L_{22}} \Rightarrow \frac{V_1}{V_2} = \frac{L_{11}}{L_{22}} \tag{4}$$

Solving (2) and (4):

$$\frac{V_1}{V_2} = \left(\frac{N_1}{N_2}\right)^2$$

Choice (1) is the answer.

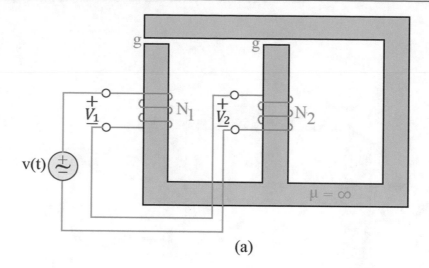

(a)

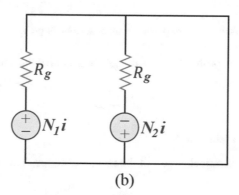

(b)

Fig. 4.2 The magnetic circuit and equivalent electrical circuit of solution of problem 4.2

4.3. Based on the information given in the problem, the core is ideal. Therefore, only the air gaps have a reluctance. Moreover, we know that the moving part is moving with the slow velocity of v.

The magnetic flux produced by the first coil can be calculated as follows:

$$\varphi = \frac{F_1}{R_{eq}} = \frac{N_1 I_1}{2R_x} = \frac{N_1 I_1}{\frac{2x}{\mu_0 A}} = \frac{N_1 I_1 \mu_0 A}{2x} \tag{1}$$

The voltage induced in the second coil can be determined as follows:

$$e_2 = -N_2 \frac{d\varphi}{dt} = -N_2 \frac{d\varphi}{dx} \times \underbrace{\frac{dx}{dt}}_{v} = -N_2 v \frac{d\varphi}{dx} \tag{2}$$

Solving (1) and (2):

$$e_2 = -N_2 v \frac{d}{dx}\left(\frac{N_1 I_1 \mu_0 A}{2x}\right) = -\frac{N_2 v N_1 I_1 \mu_0 A}{2} \frac{d}{dx}\left(\frac{1}{x}\right) = \frac{N_1 N_2 \mu_0 A I_1}{2} \frac{v}{x^2} \tag{3}$$

By assuming the following term:

$$k \triangleq \frac{N_1 N_2 \mu_0 A I_1}{2} \tag{4}$$

We have:

$$e_2 = k\frac{v}{x^2}$$

Choice (2) is the answer.

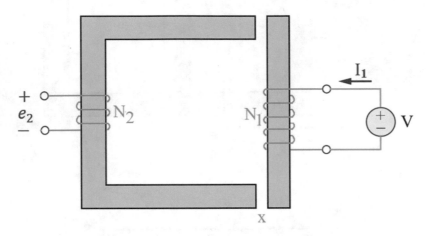

Fig. 4.3 The magnetic circuit of solution of problem 4.3

4.4. Based on the information given in the problem, the cross-sectional area of the core is the same everywhere. Moreover, we know that:

$$\mu_c \to \infty \tag{1}$$

Therefore, only the air gaps have a reluctance.

Figure 4.4.b shows the equivalent electrical circuit of the magnetic circuit. The equivalent reluctance seen by the second coil can be calculated as follows:

$$R_{eq,2} = \left(R_g \parallel R_g\right) + R_g = \frac{3}{2}R_g \tag{2}$$

The self-inductance of the second coil (L_{22}) can be calculated as follows:

$$L_{22} = \frac{N_2^2}{R_{eq,2}} = \frac{N_2^2}{\frac{3}{2}R_g} = \frac{2N_2^2}{3R_g} \tag{3}$$

To calculate the mutual inductance (L_{12}), we need to determine the amount of the magnetic flux of the second coil that flows through the first coil:

$$\varphi_2 = \frac{F_2}{R_{eq,2}} = \frac{N_2I_2}{R_{eq,2}} = \frac{N_2I_2}{\frac{3}{2}R_g} = \frac{2N_2I_2}{3R_g} \tag{4}$$

$$\Rightarrow \varphi_{12} = \frac{R_g}{R_g + R_g}\varphi_2 = \frac{N_2I_2}{3R_g} \tag{5}$$

Hence, the magnetic flux linkage of the first coil can be calculated as follows:

$$\lambda_1 = N_1\varphi_{12} = \frac{N_1N_2I_2}{3R_g} \tag{6}$$

Therefore, the mutual inductance between the coils is:

$$\Rightarrow L_{12} = \frac{\lambda_1}{I_2} = \frac{N_1N_2}{3R_g} \tag{7}$$

Solving (3) and (7):

$$\frac{L_{22}}{L_{12}} = \frac{\frac{2N_2^2}{3R_g}}{\frac{N_1N_2}{3R_g}} = \frac{2N_2}{N_1}$$

Choice (4) is the answer.

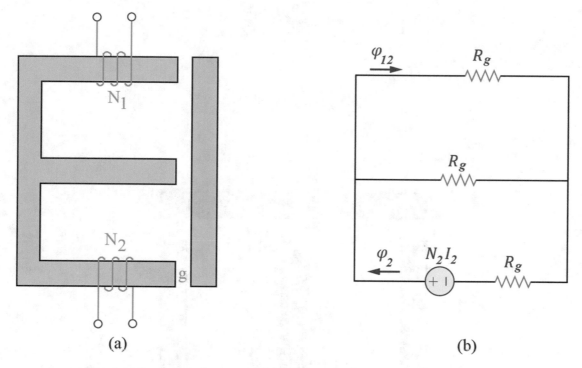

Fig. 4.4 The magnetic circuit and equivalent electrical circuit of solution of problem 4.4

4.5. Based on the information given in the problem, the cross-sectional area of the core is the same everywhere (A). Moreover, we know that:

$$\mu_c \to \infty \qquad (1)$$

Therefore, only the air gaps have a reluctance.

Figure 4.5.b shows the equivalent electrical circuit of the magnetic circuit. The equivalent reluctance seen by the first coil can be calculated as follows:

$$R_{eq,1} = R_g + \left(2R_g \parallel 2R_g\right) = 2R_g \qquad (2)$$

where R_g is the reluctance of the middle air gap and can be calculated as follows:

$$R_g = \frac{g}{\mu_0 A} \qquad (3)$$

To calculate the mutual inductance (L_{12}), we need to determine the amount of the magnetic flux of the first coil that flows through the second coil. Since both coils are placed on a common branch, all the magnetic flux of the first coil will flow through the second coil. Thus:

$$\varphi_1 = \frac{N_1 I_1}{R_{eq,1}} = \frac{N_1 I_1}{2R_g} \qquad (4)$$

$$\varphi_{21} = \frac{N_1 I_1}{2R_g} \qquad (5)$$

The magnetic flux linkage of the second coil can be calculated as follows:

$$\Rightarrow \lambda_2 = N_2\varphi_{21} = \frac{N_1N_2I_1}{2R_g} \tag{6}$$

Therefore, the mutual inductance between the coils is:

$$\Rightarrow L_{21} = \frac{\lambda_2}{I_1} = \frac{N_1N_2}{2R_g} \tag{7}$$

Solving (3) and (7):

$$\Rightarrow L_{12} = L_{21} = \frac{\mu_0 A N_1 N_2}{2g}$$

Choice (1) is the answer.

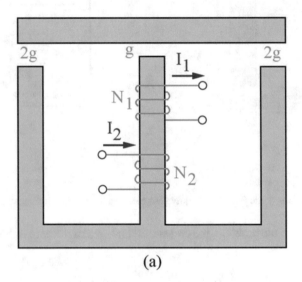

(a)

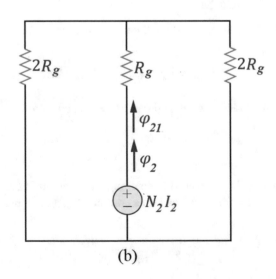

(b)

Fig. 4.5 The magnetic circuit and equivalent electrical circuit of solution of problem 4.5

4.6. Based on the information given in the problem, the reluctance of the core is zero. Therefore, only the air gaps have a reluctance. Moreover, we know that:

$$L = 10 \ mH \tag{1}$$

$$A_{middle} = 400 \ mm^2 \tag{2}$$

$$A_{outer} = 300 \ mm^2 \tag{3}$$

The reluctance of the outer and middle branches can be calculated as follows:

$$R_{middle} = \frac{g}{\mu_0 \times A_{middle}} = \frac{g}{\mu_0 \times 400 \times 10^{-6}} = \frac{g}{\mu_0} \frac{10^6}{400} \tag{4}$$

$$R_{outer} = \frac{g}{\mu_0 A_{outer}} = \frac{g}{\mu_0 \times 300 \times 10^{-6}} = \frac{g}{\mu_0} \frac{10^6}{300} \tag{5}$$

Figure 4.6.b shows the equivalent electrical circuit of the magnetic circuit. The equivalent reluctance seen by the coil can be calculated as follows:

$$R_{eq} = R_{middle} + R_{outer} \| R_{outer} = \frac{g}{\mu_0} \frac{10^6}{400} + \frac{1}{2} \times \frac{g}{\mu_0} \frac{10^6}{300} = \frac{g}{\mu_0} \frac{10^6}{240} \tag{6}$$

The inductance of the coil can be calculated as follows:

$$L = \frac{N^2}{R_{eq}} = \frac{100^2}{\frac{g}{\mu_0} \frac{10^6}{240}} = 2.4 \frac{\mu_0}{g} \tag{7}$$

Solving (1) and (7):

$$0.01 = 2.4 \frac{\mu_0}{g} \Rightarrow g = 240 \mu_0$$

Choice (4) is the answer.

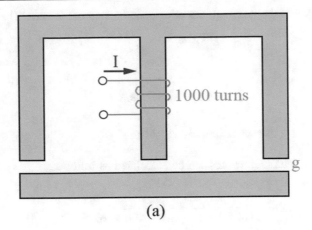

(a)

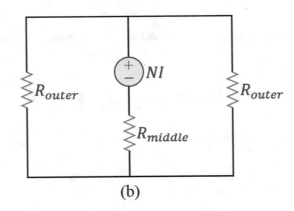

(b)

Fig. 4.6 The magnetic circuit and equivalent electrical circuit of solution of problem 4.6

4.7. Based on the information given in the problem, the reluctance of the horizontal branches is zero as well as we have:

$$N_a = N_b = N_c \triangleq N \tag{1}$$

$$A_{middle} = 2A_{outer} \tag{2}$$

As we know, the reluctance of a magnetic branch is inversely proportional to its cross-sectional area. In other words:

$$R \propto \frac{1}{A} \tag{3}$$

Solving (2) and (3):

$$R_{outer} = 2R_{middle} \triangleq 2R \tag{4}$$

Figure 4.7.b shows the equivalent electrical circuit of the magnetic system. The equivalent reluctance seen by coil "a" can be calculated as follows:

$$R_{eq,a} = R + (2R \parallel R \parallel R \parallel 2R) = \frac{4}{3}R \tag{5}$$

Thus, the self-inductance of coil "a" (L_{aa}) can be calculated as follows:

$$\Rightarrow L_{aa} = \frac{N_{aa}^2}{R_{eq,a}} = \frac{N^2}{\frac{4}{3}R} = \frac{3N^2}{4R} \tag{6}$$

To calculate the mutual inductance (L_{ab}), we need to determine the amount of the magnetic flux of coil "a" that flows through coil "b":

$$\varphi_a = \frac{F_a}{R_{eq,a}} = \frac{N_a I_a}{\frac{4}{3}R} = \frac{3NI_a}{4R} \tag{7}$$

$$\Rightarrow \varphi_{ba} = \frac{2R \parallel R \parallel 2R}{R + 2R \parallel R \parallel 2R}\varphi_a = \frac{\frac{R}{2}}{R + \frac{R}{2}}\varphi_a = \frac{\frac{R}{2}}{\frac{3R}{2}}\varphi_a = \frac{1}{3}\varphi_a = \frac{NI_a}{4R} \tag{8}$$

Hence, the magnetic flux linkage of coil "b" can be calculated as follows:

$$\Rightarrow \lambda_b = N_b \varphi_{ba} = \frac{N^2 I_a}{4R} \tag{9}$$

Therefore, the mutual inductance between coils "a" and "b" is:

$$\Rightarrow L_{ab} = L_{ba} = \frac{\lambda_b}{I_a} = \frac{N^2}{4R} \tag{10}$$

Solving (6) and (10):

$$\Rightarrow L_{aa} = 3L_{ab}$$

Choice (2) is the answer.

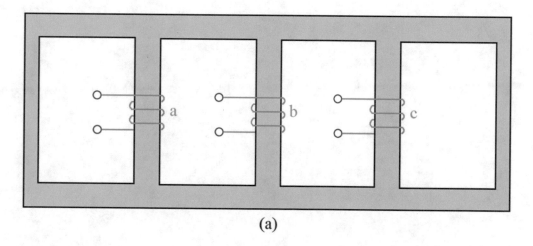

(a)

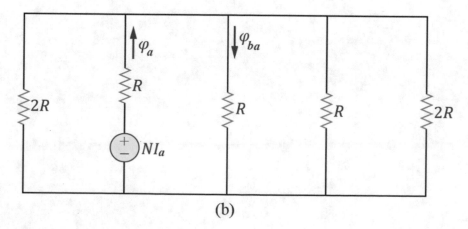

(b)

Fig. 4.7 The magnetic circuit and equivalent electrical circuit of solution of problem 4.7

Problems: Energy Loss in Magnetic Circuits

Abstract

In this chapter, the energy and power losses of magnetic circuits, including eddy current and hysteresis losses, are calculated based on the given voltage and frequency. In this chapter, the problems are categorized in different levels based on their difficulty levels (easy, normal, and hard) and calculation amounts (small, normal, and large). Additionally, the problems are ordered from the easiest problem with the smallest computations to the most difficult problems with the largest calculations.

5.1. In a magnetic core, $f = 40\ Hz$, $B_{max} = 1.2\ T$, and $P_e = 20\ W$ (eddy current loss). Calculate P_e when $f = 50\ Hz$ and $B_{max} = 1\ T$.

   Difficulty level   ○ Easy   ● Normal   ○ Hard
   Calculation amount   ● Small   ○ Normal   ○ Large

   1) 20 W
   2) 21.7 W
   3) 13.91 W
   4) 31.25 W

5.2. In a magnetic core, $f = 400\ Hz$, $B_{max} = 1.2\ T$, and $P_e = 15\ W$ (eddy current loss). Calculate P_e when $f = 480\ Hz$ and $B_{max} = 1\ T$.

   Difficulty level   ○ Easy   ● Normal   ○ Hard
   Calculation amount   ● Small   ○ Normal   ○ Large

   1) 15 W
   2) 30 W
   3) 17 W
   4) 25 W

5.3. In a magnetic core, $f = 50\ Hz$, $B_{max} = 1.2\ T$, and $P_e = 30\ W$ (eddy current loss). Calculate P_e when $f = 60\ Hz$ and $B_{max} = 1\ T$.

   Difficulty level   ○ Easy   ● Normal   ○ Hard
   Calculation amount   ● Small   ○ Normal   ○ Large

   1) 30 W
   2) 31.25 W
   3) 20 W
   4) 26.7 W

© The Author(s), under exclusive license to Springer Nature Switzerland AG 2022
M. Rahmani-Andebili, *DC Electric Machines, Electromechanical Energy Conversion Principles, and Magnetic Circuit Analysis*,
https://doi.org/10.1007/978-3-031-08863-6_5

5.4. Calculate the reduction in the value of hysteresis and eddy current losses of a single-phase transformer if its voltage and frequency decrease from 200 V and 50 Hz to 160 V and 40 Hz.

Difficulty level   ○ Easy   ● Normal   ○ Hard
Calculation amount   ● Small   ○ Normal   ○ Large

1) 32%, 36%
2) 25%, 50%
3) 20%, 36%
4) 40%, 80%

5.5. In a 60 Hz single-phase transformer, the hysteresis loss and eddy current loss are 240 W and 144 W, respectively. Calculate the core loss at 50 Hz for the constant magnetic flux.

Difficulty level   ○ Easy   ● Normal   ○ Hard
Calculation amount   ● Small   ○ Normal   ○ Large

1) 324 W
2) 314 W
3) 300 W
4) 304 W

5.6. Which one of the choices below is correct if the frequency of a transformer is doubled?

Difficulty level   ○ Easy   ● Normal   ○ Hard
Calculation amount   ○ Small   ● Normal   ○ Large

1) The hysteresis loss will increase.
2) The eddy current loss will increase.
3) The hysteresis loss will decrease.
4) The eddy current loss will decrease.

5.7. In the two no-load tests of a single-phase transformer, the results below have been obtained. Calculate the hysteresis and eddy current losses at 50 Hz.

	Voltage (V)	Frequency (Hz)	Power (W)
Test 1	220	50	2500
Test 2	110	25	675

Difficulty level   ○ Easy   ● Normal   ○ Hard
Calculation amount   ○ Small   ● Normal   ○ Large

1) $P_h = 1250\ W,\ P_e = 1250\ W$
2) $P_h = 500\ W,\ P_e = 2000\ W$
3) $P_h = 1000\ W,\ P_e = 1500\ W$
4) $P_h = 200\ W,\ P_e = 2300\ W$

5.8. In the two no-load tests of a single-phase transformer, the results below have been achieved. Calculate the ratio of eddy current loss to hysteresis loss at frequency of 50 Hz.

	Voltage (V)	Frequency (Hz)	Power (W)
Test 1	220	50	2500
Test 2	100	25	825

Difficulty level   ○ Easy   ● Normal   ○ Hard
Calculation amount   ○ Small   ● Normal   ○ Large

1) 1.875
2) 2.125
3) 2.725
4) 3.165

5.9. In the two no-load tests of a single-phase transformer, the results below have been achieved. Calculate the core loss for 50 Hz and 240 V.

	Voltage (V)	Frequency (Hz)	Power (W)
Test 1	190	40	40
Test 2	290	60	75

Difficulty level   ○ Easy   ● Normal   ○ Hard
Calculation amount   ○ Small   ● Normal   ○ Large

1) 63.5 W
2) 56.25 W
3) 53 W
4) 60 W

5.10. In a single-phase transformer with frequency of 50 Hz, the hysteresis and eddy current losses are 200 W and 100 W, respectively. With the constant magnetic flux but with the frequency of 60 Hz, calculate the core loss of the transformer.

Difficulty level   ○ Easy   ● Normal   ○ Hard
Calculation amount   ○ Small   ● Normal   ○ Large

1) 432 W
2) 384 W
3) 408 W
4) 360 W

5.11. In a no-load test of a 10 kVA transformer, we have $V = 5.5\ kV, f = 60\ Hz$, and $P = 1000\ W$. In this test, the ratio of eddy current loss to hysteresis loss is 1.5. Moreover, the hysteresis loss is proportional to $(B_{max})^{1.5}$. Calculate the no-load power if $V = 5\ kV$ and $f = 50\ Hz$.

Difficulty level   ○ Easy   ○ Normal   ● Hard
Calculation amount   ○ Small   ● Normal   ○ Large

1) 633.25 W
2) 875.70 W
3) 922.45 W
4) 1000 W

Solutions of Problems: Energy Loss in Magnetic Circuits

6

6

Abstract

In this chapter, the problems of the fifth chapter are fully solved, in detail, step-by-step, and with different methods.

6.1. Based on the information given in the problem, we have:

$$f_1 = 40 \ Hz \tag{1}$$

$$B_{max,1} = 1.2 \ T \tag{2}$$

$$P_{e,1} = 20 \ W \tag{3}$$

$$f_2 = 50 \ Hz \tag{4}$$

$$B_{max,2} = 1 \ T \tag{5}$$

The eddy current loss can be calculated by using the following relation:

$$P_e = k_e f^2 B_{max}^2 \tag{6}$$

Therefore:

$$\frac{P_{e,1}}{P_{e,2}} = \left(\frac{B_{max,1} f_1}{B_{max,2} f_2}\right)^2 \tag{7}$$

$$\Rightarrow \frac{20}{P_{e,2}} = \left(\frac{1.2 \times 40}{1 \times 50}\right)^2 \Rightarrow P_{e,2} = 21.7 \ W$$

Choice (2) is the answer.

6.2. Based on the information given in the problem, we have:

$$f_1 = 400 \ Hz \tag{1}$$

$$B_{max,1} = 1.2 \ T \tag{2}$$

$$P_{e,1} = 15 \ W \tag{3}$$

© The Author(s), under exclusive license to Springer Nature Switzerland AG 2022
M. Rahmani-Andebili, *DC Electric Machines, Electromechanical Energy Conversion Principles, and Magnetic Circuit Analysis*,
https://doi.org/10.1007/978-3-031-08863-6_6

$$f_2 = 480 \ Hz \tag{4}$$

$$B_{max,1} = 1 \ T \tag{5}$$

The eddy current loss can be calculated by using the following relation:

$$P_e = k_e f^2 B_{max}^2 \tag{6}$$

Thus:

$$\frac{P_{e,1}}{P_{e,2}} = \left(\frac{B_{max,1} f_1}{B_{max,2} f_2}\right)^2 \tag{7}$$

$$\Rightarrow \frac{15}{P_{e,2}} = \left(\frac{1.2 \times 400}{1 \times 480}\right)^2 \Rightarrow P_{e,2} = 15 \ W$$

Choice (1) is the answer.

6.3. Based on the information given in the problem, we have:

$$f_1 = 50 \ Hz \tag{1}$$

$$B_{max,1} = 1.2 \ T \tag{2}$$

$$P_{e,1} = 30 \ W \tag{3}$$

$$f_2 = 60 \ Hz \tag{4}$$

$$B_{max,2} = 1 \ T \tag{5}$$

The eddy current loss can be calculated by using the following relation:

$$P_e = k_e f^2 B_{max}^2 \tag{6}$$

Hence:

$$\frac{P_{e,1}}{P_{e,2}} = \left(\frac{B_{max,1} f_1}{B_{max,2} f_2}\right)^2$$

$$\Rightarrow \frac{30}{P_{e,2}} = \left(\frac{1.2 \times 50}{1 \times 60}\right)^2 \Rightarrow P_{e,2} = 30 \ W$$

Choice (1) is the answer.

6.4. Based on the information given in the problem, we have:

$$V_1 = 200 \ V \tag{1}$$

$$f_1 = 50 \ Hz \tag{2}$$

$$V_2 = 160 \ Hz \tag{3}$$

$$f_2 = 40 \ Hz \tag{4}$$

The hysteresis and eddy current losses can be calculated by using the following relations:

$$P_h = k_h f B_{max}^n \tag{5}$$

$$P_e = k_e f^2 B_{max}^2 \tag{6}$$

Since the value of $B_{max} = \frac{V}{f}$ is constant in the tests ($B_{max,1} = \frac{V_1}{f_1} = \frac{200}{50} = 4$ and $B_{max,2} = \frac{V_2}{f_2} = \frac{160}{40} = 4$), the new hysteresis loss (P_h) and eddy current loss (P_e) can be calculated as follows:

$$\frac{P_{h,1}}{P_{h,2}} = \frac{f_1}{f_2} = \frac{50}{40} \Rightarrow P_{h,2} = 0.8 P_{h,1} \Rightarrow 20\% \text{ Reduction}$$

$$\frac{P_{e,1}}{P_{e,2}} = \left(\frac{f_1}{f_2}\right)^2 = \left(\frac{50}{40}\right)^2 \Rightarrow P_{e,2} = 0.64 P_{e,1} \Rightarrow 36\% \text{ Reduction}$$

Choice (3) is the answer.

6.5. Based on the information given in the problem, we have:

$$f_1 = 60 \; Hz \tag{1}$$

$$P_{h,1} = 240 \; W \tag{2}$$

$$P_{e,1} = 144 \; W \tag{3}$$

$$f_2 = 50 \; Hz \tag{4}$$

$$\varphi = \text{Const.} \tag{5}$$

The hysteresis and eddy current losses can be calculated by using the following relations:

$$P_h = k_h f B_{max}^n \tag{6}$$

$$P_e = k_e f^2 B_{max}^2 \tag{7}$$

Since the value of magnetic flux is constant, the maximum value of magnetic flux density (B_{max}) will be constant. Hence, the hysteresis loss (P_h) and the eddy current loss (P_e) can be calculated as follows:

$$\frac{P_{h,1}}{P_{h,2}} = \frac{f_1}{f_2} \Rightarrow \frac{240}{P_{h2}} = \frac{60}{50} \Rightarrow P_{h,2} = 200 \; W$$

$$\frac{P_{e,1}}{P_{e,2}} = \left(\frac{f_1}{f_2}\right)^2 \Rightarrow \frac{144}{P_{e,2}} = \left(\frac{60}{50}\right)^2 \Rightarrow P_{e,2} = 100 \; W$$

Therefore, the core loss at 50 Hz is as follows:

$$P_{c,2}|_{50 \; Hz} = P_{h,2} + P_{e,2} = 200 + 100 = 300 \; W$$

Choice (3) is the answer.

6.6. The hysteresis and eddy current losses can be calculated by using the following relations:

$$P_h = k_h f B_{max}^n \xrightarrow{B_{max} = \frac{V}{f}} P_h = f^{1-n} V^n$$

$$P_e = k_e f^2 B_{max}^2 \xrightarrow{B_{max} = \frac{V}{f}} P_h = V^2$$

Therefore, if only the frequency is doubled, the hysteresis loss will decrease but the eddy current loss will not change.

Choice (3) is the answer.

6.7. Based on the information given in the problem, we have:

	Voltage (V)	Frequency (Hz)	Power (W)
Test 1	220	50	2500
Test 2	110	25	675

The hysteresis and eddy current losses can be calculated by using the following relations:

$$P_h = k_h f B_{max}^n \tag{1}$$

$$P_e = k_e f^2 B_{max}^2 \tag{2}$$

Since the value of $B_{max} = \frac{V}{f}$ is constant in the tests, the core loss of the transformer, including the hysteresis loss (P_h) and the eddy current loss (P_e), can be calculated as follows:

$$P_c = P_h + P_e = A_h f + A_e f^2 \tag{3}$$

Applying (3) for both tests:

$$\Rightarrow \begin{cases} \text{Test 1} : 2500 = A_h \times 50 + A_e \times 50^2 \\ \text{Test 2} : 675 = A_h \times 25 + A_e \times 25^2 \end{cases} \Rightarrow A_h = 4, \quad A_e = 0.92 \tag{4}$$

Therefore, the hysteresis and eddy current losses at 50 Hz are as follows:

$$P_h|_{50\ Hz} = 4 \times 50 \Rightarrow P_h|_{50\ Hz} = 200 \text{ W}$$

$$P_e|_{50\ Hz} = 0.92 \times 50^2 \Rightarrow P_e|_{50\ Hz} = 2300 \text{ W}$$

Choice (4) is the answer.

6.8. Based on the information given in the problem, we have:

	Voltage (V)	Frequency (Hz)	Power (W)
Test 1	220	50	2500
Test 2	100	25	825

The hysteresis and eddy current losses can be calculated by using the following relations:

$$P_h = k_h f B_{max}^n \tag{1}$$

$$P_e = k_e f^2 B_{max}^2 \tag{2}$$

Since the value of $B_{max} = \frac{V}{f}$ is almost constant in the tests, the core loss of the transformer, including the hysteresis loss (P_h) and eddy current loss (P_e), can be calculated as follows:

$$P_c = P_h + P_e = A_h f + A_e f^2 \tag{3}$$

Applying (3) for both tests:

$$\Rightarrow \begin{cases} \text{Test 1} : 2500 = A_h \times 50 + A_e \times 2500 \\ \text{Test 2} : 825 = A_h \times 25 + A_e \times 625 \end{cases} \Rightarrow A_h = 16, \quad A_e = 0.68 \tag{4}$$

Therefore, the hysteresis and eddy current losses at 50 Hz are as follows:

$$P_h|_{50\ Hz} = 16 \times 50 = 800 \text{ W} \tag{5}$$

$$P_e|_{50\ Hz} = 0.68 \times 50^2 = 1700 \text{ W} \tag{6}$$

$$\Rightarrow \frac{P_e}{P_h}\Big|_{50\ Hz} = 2.125$$

Choice (2) is the answer.

6.9. Based on the information given in the problem, we have:

	Voltage (V)	Frequency (Hz)	Power (W)
Test 1	190	40	40
Test 2	290	60	75

The hysteresis and eddy current losses can be calculated by using the following relations:

$$P_h = k_h f B_{max}^n \tag{1}$$

$$P_e = k_e f^2 B_{max}^2 \tag{2}$$

Since the value of $B_{max} = \frac{V}{f}$ is almost constant in the tests, the core loss of the transformer, including the hysteresis loss (P_h) and eddy current loss (P_e), can be calculated as follows:

$$P_c = P_h + P_e = A_h f + A_e f^2 \tag{3}$$

Applying (3) for both tests:

$$\Rightarrow \begin{cases} \text{Test 1} : 40 = A_h \times 40 + A_e \times 40^2 \\ \text{Test 2} : 75 = A_h \times 60 + A_e \times 60^2 \end{cases} \Rightarrow A_h = \frac{1}{2}, \quad A_e = \frac{1}{80} \tag{4}$$

Therefore, the core loss at 50 Hz is as follows:

$$P_c = A_h f + A_e f^2 = \frac{1}{2} \times 50 + \frac{1}{80} \times 50^2 \Rightarrow P_c = 56.25 \text{ W}$$

Choice (2) is the answer.

6.10. Based on the information given in the problem, we have:

$$f_1 = 50 \ Hz \tag{1}$$

$$P_{h,1} = 200 \ W \tag{2}$$

$$P_{e,1} = 100 \ W \tag{3}$$

$$\varphi = \text{Const.} \tag{4}$$

The hysteresis and eddy current losses can be calculated by using the following relations:

$$P_h = k_h f B_{max}^n \tag{5}$$

$$P_e = k_e f^2 B_{max}^2 \tag{6}$$

Since the value of magnetic flux is constant, the maximum value of magnetic flux density (B_{max}) will be constant. Hence, the hysteresis loss (P_h) and the eddy current loss (P_e) can be calculated as follows:

$$\frac{P_{h,1}}{P_{h,2}} = \frac{f_1}{f_2} \Rightarrow \frac{200}{P_{h,2}} = \frac{50}{60} \Rightarrow P_{h,2} = 240 \ W \tag{7}$$

$$\frac{P_{e,1}}{P_{e,2}} = \left(\frac{f_1}{f_2}\right)^2 \Rightarrow \frac{100}{P_{e,2}} = \left(\frac{50}{60}\right)^2 \Rightarrow P_{e,2} = 144 \ W \tag{8}$$

Therefore, the core loss at 60 Hz is as follows:

$$\Rightarrow P_c = 240 + 144 = 384 \ W$$

Choice (2) is the answer.

6.11. Based on the information given in the problem, we have:

$$V_1 = 5.5 \ kV \tag{1}$$

$$f_1 = 60 \ Hz \tag{2}$$

$$P_{NL,1} = P_{c,1} = 1000 \ W \tag{3}$$

$$\frac{P_e}{P_h} = 1.5 \tag{4}$$

$$P_h \propto (B_{max})^{1.5} \tag{5}$$

$$V_1 = 5 \ kV \tag{6}$$

$$f_1 = 50 \ Hz \tag{7}$$

Solving (3) and (4):

$$\begin{cases} P_{h,1} + P_{e,1} = 1000 \ W \\ P_{e,1} = 1.5 P_{h,1} \end{cases} \Rightarrow P_{h,1} = 400 \ W, \quad P_{e,1} = 600 \ W \tag{8}$$

The new hysteresis loss can be calculated as follows:

$$P_h = k_h f B_{max}^n \xrightarrow{B_{max} = \left(\frac{V}{f}\right)^{1.5}} P_h = f^{-0.5} V^{1.5} \tag{9}$$

$$\frac{P_{h,1}}{P_{h,2}} = \left(\frac{V_1}{V_2}\right)^{1.5} \left(\frac{f_1}{f_2}\right)^{-0.5} \Rightarrow \frac{400}{P_{h,2}} = \left(\frac{5.5}{5}\right)^{1.5} \left(\frac{60}{50}\right)^{-0.5} \Rightarrow P_{h,2} = 379.8 \ W \tag{10}$$

The new eddy current loss can be calculated as follows:

$$P_e = k_e f^2 B_{max}^2 \xrightarrow{\quad B_{max} = \frac{V}{f} \quad} P_h = V^2 \tag{11}$$

$$\frac{P_{e,1}}{P_{e,1}} = \left(\frac{V_1}{V_2}\right)^2 \Rightarrow \frac{600}{P_{e,2}} = \left(\frac{5.5}{5}\right)^2 \Rightarrow P_{e,2} = 495.9 \; W \tag{12}$$

Therefore, the new core loss is as follows:

$$\Rightarrow P_{c,2} = 379.8 + 495.9 = 875.7 \; W$$

Choice (2) is the answer.

Abstract

In this chapter, the problems concerned with the principles of electromechanical energy conversion are presented. Herein, the amounts of energy and co-energy saved in the magnetic circuits are calculated. Moreover, the electromagnetic force and torque exerted on the electromechanical relay are determined. In this chapter, the problems are categorized in different levels based on their difficulty levels (easy, normal, and hard) and calculation amounts (small, normal, and large). Additionally, the problems are ordered from the easiest problem with the smallest computations to the most difficult problems with the largest calculations.

7.1. The magnetic flux linkage-current ($\lambda - i$) of a magnetic circuit is shown in Fig. 7.1. For point b, calculate the value of energy (W_f) and co-energy (W_f') saved in the magnetic circuit.

Difficulty level	● Easy	○ Normal	○ Hard
Calculation amount	● Small	○ Normal	○ Large

1) 1.5 J, 1.5 J
2) 1.25 J, 1.75 J
3) 1.1 J, 1.9 J
4) 0.8 J, 1.2 J

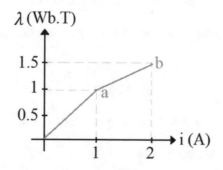

Fig. 7.1 The graph of problem 7.1

7.2. The magnetic flux linkage-current ($\lambda - i$) of a magnetic circuit is illustrated in Fig. 7.2. Calculate the ratio of energy saved in the magnetic circuit from $i = 0$ to $i = 1$ ($W_{f, i = 0 \rightarrow 1}$) to the energy saved in the circuit from $i = 1$ to $i = 3$ ($W_{f, i = 1 \rightarrow 3}$).

Difficulty level	● Easy	○ Normal	○ Hard
Calculation amount	● Small	○ Normal	○ Large

1) 4
2) 0.25
3) 0.5
4) 2

© The Author(s), under exclusive license to Springer Nature Switzerland AG 2022
M. Rahmani-Andebili, *DC Electric Machines, Electromechanical Energy Conversion Principles, and Magnetic Circuit Analysis*,
https://doi.org/10.1007/978-3-031-08863-6_7

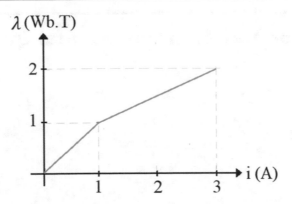

Fig. 7.2 The graph of problem 7.2

7.3. In an electromechanical system, the relation between the magnetic flux linkage and the current is $\lambda = \sqrt{i}$. Calculate the amount of electromagnetic force exerted on its moving part for $i = 2\ A$.

Difficulty level   ○ Easy   ● Normal   ○ Hard
Calculation amount   ● Small   ○ Normal   ○ Large

1) 1.88 N
2) 2.88 N
3) 0.66 N
4) 1.33 N

7.4. Which one of the following choices is correct about an electromagnetic force or torque in an electromechanical system?

Difficulty level   ○ Easy   ● Normal   ○ Hard
Calculation amount   ● Small   ○ Normal   ○ Large

1) In a constant magnetomotive force (mmf), it acts in a way that increases the energy and decreases the co-energy saved in the system.
2) In a constant current, it acts in a way that increases both energy and co-energy saved in the system.
3) In a constant flux, it acts in a way that increases energy saved in the system.
4) It acts in a way that increases the magnetic reluctance and inductance.

7.5. In an electromechanical system, the relation between the magnetic flux linkage and the current is as follows:

$$i = \left(\frac{\lambda}{x}\right)^2$$

Calculate the electromagnetic force exerted on its moving part for $i = 1\ A$.

Difficulty level   ○ Easy   ● Normal   ○ Hard
Calculation amount   ○ Small   ● Normal   ○ Large

1) $\frac{1}{3}\ N$
2) $\frac{3}{4}\ N$
3) $\frac{3}{2}\ N$
4) $\frac{2}{3}\ N$

7.6. In a magnetic core, the relation between the magnetic flux linkage and the current is as follows:

$$\lambda = 4i^{\frac{1}{3}}$$

Calculate the energy saved in the magnetic circuit when $i = 8\ A$.

Difficulty level   ○ Easy   ● Normal   ○ Hard
Calculation amount   ○ Small   ● Normal   ○ Large

1) 8 J
2) 4 J
3) 16 J
4) 12 J

7.7. In a magnetic circuit, the relation between the magnetic flux linkage and the current is as follows:

$$\lambda = 0.3\sqrt{i}$$

Calculate the energy saved in the magnetic circuit when $i = 4\ A$.

Difficulty level   ○ Easy   ● Normal   ○ Hard
Calculation amount   ○ Small   ● Normal   ○ Large
1) 1.2 J
2) 2.4 J
3) 1.6 J
4) 0.8 J

7.8. In the magnetic circuit of an electromechanical relay, shown in Fig. 7.3, the relation between the magnetic flux linkage and the current is as follows:

$$\lambda = \frac{4}{(x+1)}\left[i^{\frac{1}{2}} + i^{\frac{1}{3}}\right]$$

Calculate the electromagnetic force exerted on the moving part of the relay for $i = 1\ A$ and $x = 1\ m$.

Difficulty level   ○ Easy   ● Normal   ○ Hard
Calculation amount   ○ Small   ● Normal   ○ Large
1) $-\frac{17}{12}\ N$
2) $-\frac{12}{17}\ N$
3) $\frac{17}{12}\ N$
4) $\frac{12}{17}\ N$

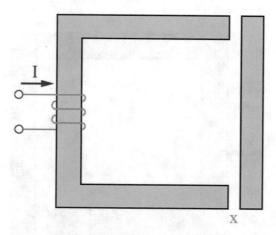

Fig. 7.3 The magnetic circuit of problem 7.8

7.9. Figure 7.4 shows an electromechanical relay. Parametrically determine the electromagnetic force exerted on the moving part of the relay if $i_s(t) = I_m \cos \omega t$ and $\mu_r \rightarrow \infty$.

Difficulty level   ○ Easy   ● Normal   ○ Hard
Calculation amount   ○ Small   ● Normal   ○ Large

1) $-\frac{1}{2}\frac{N^2 I_m^2 \mu_0 A}{g^2}$

2) $-\frac{1}{2}\frac{N^2 I_m \mu_0 A}{g}$

3) $-\frac{N^2 I_m \mu_0 A}{g}$

4) $-\frac{1}{4}\frac{N^2 I_m^2 \mu_0 A}{g^2}$

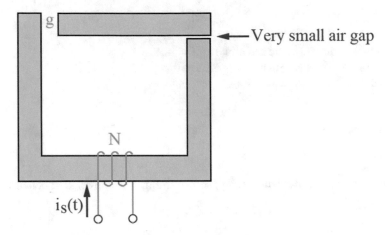

Fig. 7.4 The magnetic circuit of problem 7.9

7.10. Figure 7.5 shows an electromechanical relay. Parametrically determine the maximum value of the average electromagnetic force exerted on the moving part of the relay if $i_s(t) = I_0 + I_0 \sin \omega t$ and $\mu_c \to \infty$.

Difficulty level   ○ Easy   ● Normal   ○ Hard
Calculation amount   ○ Small   ● Normal   ○ Large

1) $\frac{1}{4}N^2 \mu_0 A \left(\frac{I_0}{x}\right)^2$

2) $\frac{3}{5}N^2 \mu_0 A \left(\frac{I_0}{x}\right)^2$

3) $\frac{3}{4}N^2 \mu_0 A \left(\frac{I_0}{x}\right)^2$

4) $\frac{3}{8}N^2 \mu_0 A \left(\frac{I_0}{x}\right)^2$

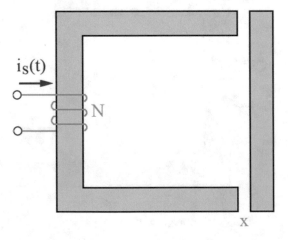

Fig. 7.5 The magnetic circuit of problem 7.10

7.11. Figure 7.6 shows an electromechanical relay. Calculate the value of current that must flow through the coil so that an electromagnetic force with the magnitude of 11.25 N is exerted on the wood when the air gap is 1 cm. In this problem, assume that $\mu_c \to \infty$, $\mu_0 = 10^{-6}$ H/m, and $A = 20$ cm^2.

Difficulty level   ○ Easy   ● Normal   ○ Hard
Calculation amount   ○ Small   ● Normal   ○ Large

1) 1.5 A
2) 2.5 A
3) 1 A
4) 2 A

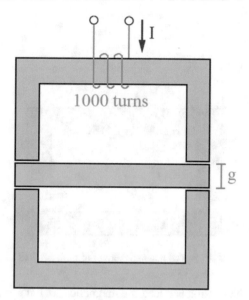

Fig. 7.6 The magnetic circuit of problem 7.11

7.12. In an electromechanical relay shown in Fig. 7.7, the magnetic core is ideal. If the magnetic flux in the air gap is about 1 mWb, calculate the energy saved in the magnetic circuit. Herein, the length of air gap is 10 mm and the cross-sectional area of the core is 25 cm².

Difficulty level   ○ Easy   ● Normal   ○ Hard
Calculation amount   ○ Small   ● Normal   ○ Large
1) 0.60 J
2) 3.18 J
3) 1.20 J
4) 0.30 J

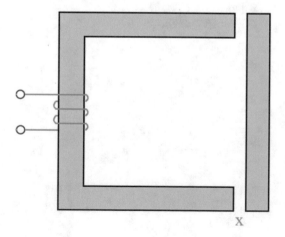

Fig. 7.7 The magnetic circuit of problem 7.12

7.13. In an electromechanical system, shown in Fig. 7.8, the relation between the magnetic flux linkage and the current is as follows:

$$i = \lambda^{\frac{1}{2}} + 20\lambda(x - 0.1)^2$$

Calculate the electromagnetic force exerted on its moving part when $x = 0.05$ m.

Difficulty level   ○ Easy   ● Normal   ○ Hard
Calculation amount   ○ Small   ● Normal   ○ Large

1) λ^2

2) 0.5λ

3) $0.5\lambda^{-0.5} + 0.05\lambda$

4) $\frac{2}{3}\lambda^{1.5} + 0.025\lambda^2$

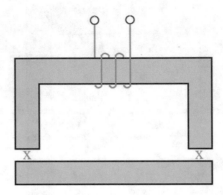

Fig. 7.8 The magnetic circuit of problem 7.13

7.14. In an electromechanical system, the relation between the magnetic flux linkage and the current is $\lambda = 0.66i^{\frac{1}{3}}$. Calculate the energy saved in the system for $i = 1\ A$.

Difficulty level   ○ Easy   ● Normal   ○ Hard
Calculation amount   ○ Small   ● Normal   ○ Large

1) $0.165\ J$

2) $0.66\ J$

3) $0.33\ J$

4) $1\ J$

7.15. In an electromechanical relay, shown in Fig. 7.9, the magnetic core is ideal, the cross-sectional area of the core is $10\ cm^2$, and the relative permeability of the air gap is $\frac{1}{\mu_0} = 800,000$. If the voltage applied to the 1000-turn coli is $v(t) = 20\sin 100t$, how much is the magnitude of electromagnetic force exerted on the moving part?

Difficulty level   ○ Easy   ● Normal   ○ Hard
Calculation amount   ○ Small   ● Normal   ○ Large

1) $32\ N$

2) $16\ N$

3) $8\ N$

4) $24\ N$

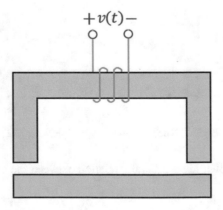

Fig. 7.9 The magnetic circuit of problem 7.15

7.16. In an electromechanical system, the relation between the magnetic flux linkage and the current is as follows:

$$\lambda = \sqrt{\frac{i}{x}}$$

Calculate the electromagnetic force exerted on the moving part for $i = 1\ A$ and $x = 1\ m$.

Difficulty level   ○ Easy   ● Normal   ○ Hard
Calculation amount   ○ Small   ● Normal   ○ Large

1) $\frac{2}{3}\ N$

2) $-\frac{2}{3}\ N$

3) $\frac{1}{3}\ N$

4) $-\frac{1}{3}\ N$

7.17. In an electromechanical relay, shown in Fig. 7.10, the magnetic core is ideal, the cross-sectional area of the core is $10\ cm^2$, and the number of turns of the coil is 1000. Calculate the magnitude of electromagnetic force exerted on the moving part if $i = 10\ A$ and $g = 5\ mm$?

Difficulty level   ○ Easy   ● Normal   ○ Hard
Calculation amount   ○ Small   ● Normal   ○ Large

1) $2\pi\ N$

2) $8\pi\ N$

3) $4\pi\ N$

4) $3\pi\ N$

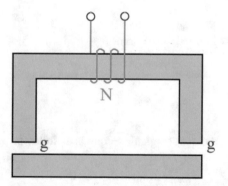

Fig. 7.10   The magnetic circuit of problem 7.17

7.18. In an electric machine that includes two exciting windings, the mutual inductance is $L_{sr}(\theta) = 0.04 \cos \theta$. Calculate the maximum value of average electromagnetic torque if the current of $10\sqrt{2} \sin \omega t$ flows through the windings of stator and rotor.

Difficulty level   ○ Easy   ● Normal   ○ Hard
Calculation amount   ○ Small   ● Normal   ○ Large

1) $3\ N.\ m$

2) $2\ N.\ m$

3) $1\ N.\ m$

4) $4\ N.\ m$

7.19. In an electric machine, the energy saved in the magnetic circuit is as follows:

$$W = 5i_s^2 - 2i_r i_s \cos \theta + 0.04 i_r^2 \cos 2\theta$$

where θ is the angle between the axes of rotor and stator. Which one of the following choices is correct?

Difficulty level   ○ Easy   ● Normal   ○ Hard
Calculation amount   ○ Small   ● Normal   ○ Large

1) The stator has salient poles, but the rotor is cylindrical.
2) The rotor has salient poles, but the stator is cylindrical.
3) Both stator and rotor have salient poles.
4) Both stator and rotor are cylindrical.

7.20. The inductances of a two-part electromechanical energy convertor are $L_{11} = 0.1\ H$, $L_{22} = 0.15\ H$, $L_{12} = 0.04 \cos \theta\ H$, where θ is the angle between the axes of the windings. If $i(t) = \sin \omega t\ A$ flows through both windings, calculate the maximum value of torque of the convertor.

Difficulty level   ○ Easy   ● Normal   ○ Hard
Calculation amount   ○ Small   ● Normal   ○ Large

1) 0.04 N. m
2) 0.02 N. m
3) 0.25 N. m
4) 0.1 N. m

7.21. In an electric machine, the inductance matrix is as follows, where θ is the angle between the axes of rotor and stator:

$$L = \begin{bmatrix} 1 & 0.5 \cos \theta \\ 0.5 \cos \theta & 1.5 \end{bmatrix} H$$

If the currents flowing through the windings of stator and rotor are 5 A and 10 A, respectively, calculate the average electromagnetic torque for $\theta = 90°$.

Difficulty level   ○ Easy   ● Normal   ○ Hard
Calculation amount   ○ Small   ● Normal   ○ Large

1) −12.5 N. m
2) −25 N. m
3) 12.5 N. m
4) −6.25 N. m

7.22. In an electric machine, the developed electromagnetic torque is as follows:

$$T_{ave} = 2 i_s i_r \sin \theta - 0.04 i_r^2 \sin 2\theta$$

where θ is the angle between the axes of rotor and stator. Which one of the following choices is correct?

Difficulty level   ○ Easy   ● Normal   ○ Hard
Calculation amount   ○ Small   ● Normal   ○ Large

1) Both stator and rotor are cylindrical.
2) The rotor has salient poles, but the stator is cylindrical.
3) Both stator and rotor have salient poles.
4) The stator has salient poles, but the rotor is cylindrical.

7.23. In an electromagnetic relay that includes two exciting windings, the inductance matrix is as follows:

$$L = \begin{bmatrix} 1 + x & 1 - x \\ 1 - x & 2 + x \end{bmatrix}$$

Calculate the repulsive electromagnetic force developed for $|I_1| = 1\ A$ and $|I_2| = 2\ A$.

Difficulty level   ○ Easy   ○ Normal   ● Hard
Calculation amount   ○ Small   ● Normal   ○ Large

1) 0.5 N
2) 4.5 N
3) 2.5 N
4) 3.5 N

7.24. Figure 7.11 shows an electromechanical system that includes an ideal core. If the air gap decreases from x_1 to x_2 with an infinite speed, calculate the ratio of I_1 (when the air gap is x_1) to I_2 (when the air gap is x_2).

Difficulty level   ○ Easy   ○ Normal   ● Hard
Calculation amount   ○ Small   ● Normal   ○ Large

1) $\frac{I_1}{I_2} = \frac{x_1}{x_2}$

2) $\frac{I_2}{I_1} = \frac{\sqrt{x_1}+1}{\sqrt{x_2}+1}$

3) $\frac{I_2}{I_1} = \sqrt{x_1 x_2}$

4) $\frac{I_2}{I_1} = \sqrt{\frac{x_1}{x_2}}$

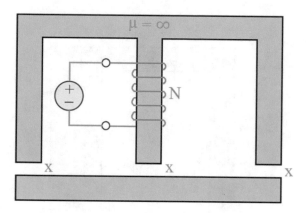

Fig. 7.11 The magnetic circuit of problem 7.24

7.25. In an electromechanical system, shown in Fig. 7.12, parametrically calculate the electromagnetic torque exerted on the moving part.

Difficulty level   ○ Easy   ○ Normal   ● Hard
Calculation amount   ○ Small   ● Normal   ○ Large

1) $-k\left(\frac{I}{\theta}\right)^2$

2) $k\frac{I^2}{\theta}$

3) $k\left(\frac{I}{\theta}\right)^2$

4) $-k\frac{I^2}{\theta}$

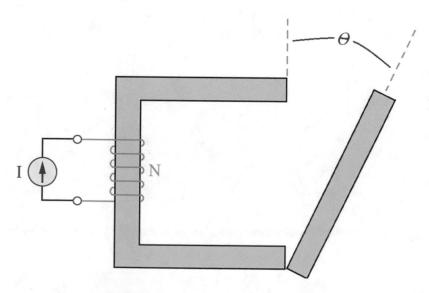

Fig. 7.12 The magnetic circuit of problem 7.25

7.26. In an electromechanical system, shown in Fig. 7.13, the magnetic flux density in the air gap of g_y is 0.4 T. The cross-sectional areas in parts A and B are 2 cm^2 and 4 cm^2, respectively. Calculate F_x and F_y if the core is assumed ideal and there is no flux leakage in the system.

Difficulty level   ○ Easy   ○ Normal   ● Hard
Calculation amount   ○ Small   ● Normal   ○ Large

1) $F_x = F_y = \frac{80}{\pi} N$

2) $F_x = F_y = \frac{160}{\pi} N$

3) $F_x = \frac{80}{\pi} N, F_y = \frac{160}{\pi} N$

4) $F_x = \frac{160}{\pi} N, F_y = \frac{80}{\pi} N$

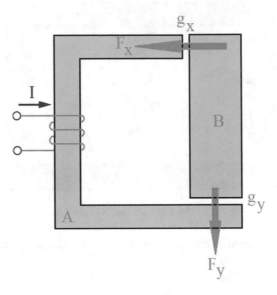

Fig. 7.13 The magnetic circuit of problem 7.26

7.27. A copper plate with the thickness of d mm has been surrounded by two U-shape magnets with the cross-sectional area of A cm^2 and the relative permeability of $\mu_r \gg 1$, as is shown in Fig. 7.14. Herein, assume that there is no flux leakage in the system. Parametrically calculate the absolute value of total electromagnetic force exerted on each side of the copper plate.

Difficulty level   ○ Easy   ○ Normal   ● Hard
Calculation amount   ○ Small   ● Normal   ○ Large

1) $\mu_0 A \frac{(N_1 - N_2)^2 I^2}{4d^2}$

2) $\mu_0 A \frac{(N_1^2 - N_2^2) I^2}{2d^2}$

3) $\mu_0 A \frac{(N_1^2 - N_2^2) I^2}{4d^2}$

4) 0

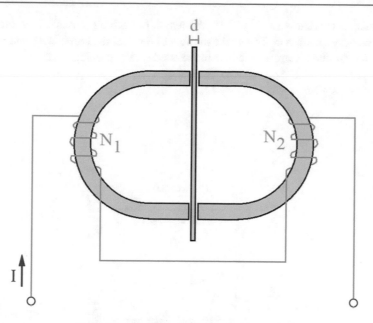

Fig. 7.14 The magnetic circuit of problem 7.27

7.28. In the electromechanical system shown in Fig. 7.15, the moving part is locked. If it is let to horizontally move, which one of the choices below is correct about the final status of the system? Herein, the thickness of the core is constant everywhere.

Difficulty level   ○ Easy   ○ Normal   ● Hard
Calculation amount   ○ Small   ● Normal   ○ Large

1) The moving part will stop in a position that the volume of the air gap in both sides is equal.
2) The moving part will stop in a position that the energy saved in both sides is equal.
3) Since the ampere-turn of the coils is equal, no electromagnetic force will be exerted on the moving part.
4) The moving part will move to one side and attach to the fixed part.

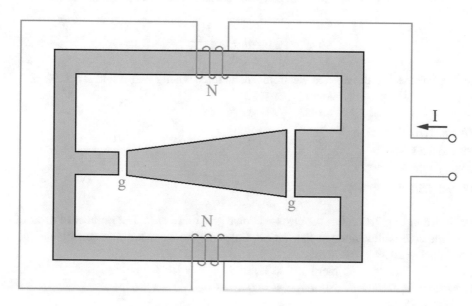

Fig. 7.15 The magnetic circuit of problem 7.28

7.29. In the electromechanical relay, shown in Fig. 7.16, the length of each air gap is 1 *mm*, the magnetic core is ideal, the depth of core is 10 *cm* everywhere, $a = 2.5$ *cm*, the number of turns of the coil is 200, and the effective value of current of the coil is 1 A. Calculate the electromagnetic force exerted on the moving part.

Difficulty level   ○ Easy   ○ Normal   ● Hard
Calculation amount   ○ Small   ● Normal   ○ Large

1) -20π *N*
2) 40π *N*
3) -2π *N*
4) 4π *N*

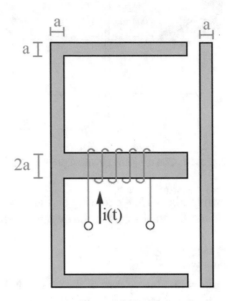

Fig. 7.16 The magnetic circuit of problem 7.29

7.30. The energy saved in the magnetic circuit of an electromechanical energy convertor is as follows:

$$W_f(\lambda, x) = \frac{\lambda^3}{0.1 - x}$$

Calculate the flux linkage and the exerted electromagnetic force when $i = 3$ *A* and $x = 0.02$ *m*.

Difficulty level   ○ Easy   ○ Normal   ● Hard
Calculation amount   ○ Small   ● Normal   ○ Large

1) $0.1\sqrt{2}$ *Wb* and 1.76 *N*
2) $0.1\sqrt{2}$ *Wb* and 3.53 *N*
3) $0.2\sqrt{2}$ *Wb* and 1.76 *N*
4) $0.2\sqrt{2}$ *Wb* and 3.53 *N*

7.31. The iron core shown in Fig. 7.17.a has the cross-sectional area of 16 *cm²*. The number of turns of the coil is 210. The $B - H$ curve of the core is illustrated in Fig. 7.17.b. Calculate the magnetic flux density and the energy saved in the magnetic circuit for $I = 2A$.

Difficulty level   ○ Easy   ○ Normal   ● Hard
Calculation amount   ○ Small   ○ Normal   ● Large

1) $B = 1.5$ *T*, $W = 0.75$ *J*
2) $B = 1.5$ *T*, $W = 0.056$ *J*
3) $B = 1.5$ *T*, $W = 0.252$ *J*
4) $B = 1$ *T*, $W = 0.056$ *J*

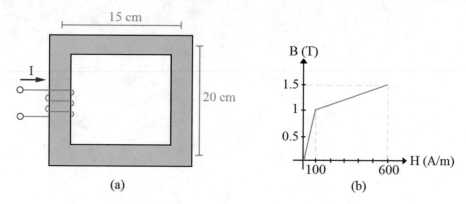

Fig. 7.17 The magnetic circuit of problem 7.31

Solutions of Problems: Electromechanical Energy Conversion

8

Abstract

In this chapter, the problems of the seventh chapter are fully solved, in detail, step-by-step, and with different methods.

8.1. The area above and under the curve of magnetic flux linkage-current ($\lambda - i$) presents the value of energy (W_f) and co-energy (W'_f) saved in the magnetic circuit, respectively.

Therefore, for the given magnetic system, we have:

$$W_f = \left(\frac{1}{2} \times 1\right) + \left(\frac{1+2}{2} \times 0.5\right) \Rightarrow W_f = 1.25\, J$$

$$W'_f = \left(\frac{1}{2} \times 1\right) + \left(\frac{1+1.5}{2} \times 1\right) \Rightarrow W'_f = 1.75\, J$$

Choice (2) is the answer.

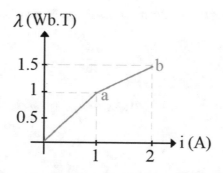

Fig. 8.1 The graph of solution of problem 8.1

8.2. The area above the curve of magnetic flux linkage-current ($\lambda - i$) presents the value of energy saved in the magnetic circuit (W_f).

Therefore, for the given magnetic system, we can write:

$$\frac{W_{f,i=0\to1}}{W_{f,i=1\to3}} = \frac{\text{Area of Triangle}}{\text{Area of Trapezius}} = \frac{S_1}{S_2} = \frac{\frac{1\times1}{2}}{\frac{(1+3)\times1}{2}}$$

$$\Rightarrow \frac{W_{f,i=0\to1}}{W_{f,i=1\to3}} = 0.25$$

Choice (2) is the answer.

© The Author(s), under exclusive license to Springer Nature Switzerland AG 2022
M. Rahmani-Andebili, *DC Electric Machines, Electromechanical Energy Conversion Principles, and Magnetic Circuit Analysis*,
https://doi.org/10.1007/978-3-031-08863-6_8

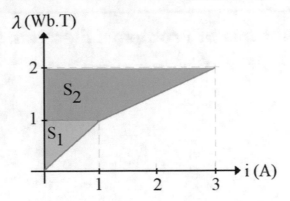

Fig. 8.2 The graph of solution of problem 8.2

8.3. Based on the information given in the problem, the relation below exists between the magnetic flux linkage and the current of the magnetic circuit:

$$\lambda = \sqrt{i} \tag{1}$$

$$i = 2\,A \tag{2}$$

The co-energy saved in the electromagnetic system can be calculated as follows:

$$W_f' = \int \lambda di \tag{3}$$

$$\Rightarrow W_f' = \int \sqrt{i}xdi = \frac{2}{3}xi^{\frac{3}{2}} \tag{4}$$

The average electromagnetic force exerted on the moving part can be calculated as follows:

$$F_{ave} = \frac{\partial W_f'}{\partial x} \tag{5}$$

$$\Rightarrow F_{ave} = \frac{2}{3}i^{\frac{3}{2}}\big|_{i=2A} \Rightarrow F_{ave} = 1.88\,N$$

Choice (1) is the answer.

8.4. An electromagnetic force or torque in an electromechanical system acts in a way that, in a constant current, increases both energy and co-energy saved in the system. Choice (2) is the answer.

8.5. Based on the information given in the problem, the relation below exists between the magnetic flux linkage and the current of the magnetic circuit:

$$i = \left(\frac{\lambda}{x}\right)^2 \tag{1}$$

$$\Rightarrow \lambda = x\sqrt{i} \tag{2}$$

Moreover, we have:

$$i = 1\,A \tag{3}$$

Method 1: From (2), we see that the relation between the magnetic flux linkage and the current is in the form of $\lambda = \lambda(i, x)$; therefore, the current is an independent variable, and we can determine the co-energy saved in the magnetic circuit as follows:

$$W'_f = \int \lambda di \tag{4}$$

Solving (2) and (4):

$$W'_f = \int x\sqrt{i}\, di = \frac{2}{3}xi^{\frac{3}{2}} \tag{5}$$

The average electromagnetic force exerted on the moving part can be calculated as follows:

$$F_{ave} = \frac{\partial W'_f}{\partial x} \tag{6}$$

Solving (5) and (6):

$$F_{ave} = \frac{2}{3}i^{\frac{3}{2}}\big|_{i=1\ \text{A}} = \frac{2}{3}N$$

Method 2: From (1), we see that the relation between the magnetic flux linkage and the current is in the form of $i = i(\lambda, x)$; therefore, the flux linkage is an independent variable, and we can determine the energy saved in the magnetic circuit as follows:

$$W_f = \int i d\lambda \tag{7}$$

Solving (1) and (7):

$$W_f = \int \frac{\lambda^2}{x^2} d\lambda = \frac{1}{x^2} \frac{\lambda^3}{3} \tag{8}$$

The average electromagnetic force exerted on the moving part can be calculated as follows:

$$F_{ave} = -\frac{\partial W_f}{\partial x} \tag{9}$$

Solving (8) and (9):

$$F_{ave} = \frac{2}{3} \frac{\lambda^3}{x^3} \tag{10}$$

Solving (2) and (3):

$$\lambda = x \tag{11}$$

Solving (10) and (11):

$$F_{ave} = \frac{2}{3}N$$

Choice (4) is the answer.

8.6. Based on the information given in the problem, the relation below exits between the magnetic flux linkage and current in the magnetic core:

$$\lambda = 4i^{\frac{1}{3}} \tag{1}$$

Moreover, we have:

$$i = 8 \, A \tag{2}$$

From Eq. (1), we have:

$$\xRightarrow{d} d\lambda = \frac{4}{3} i^{-\frac{2}{3}} di \tag{3}$$

The energy saved in the magnetic circuit can be calculated as follows:

$$W_f = \int i d\lambda \tag{4}$$

Solving (3) and (4):

$$\Rightarrow W_f = \int i\left(\frac{4}{3} i^{-\frac{2}{3}} di\right) = \int \frac{4}{3} i^{\frac{1}{3}} di$$

The energy saved in the magnetic circuit for $i = 8 \, A$ can be calculated as follows:

$$\Rightarrow W_f = \int_0^8 \frac{4}{3} i^{\frac{1}{3}} di = i^{\frac{4}{3}} \Big|_0^8 = 16 \, J$$

Choice (3) is the answer.

8.7. Based on the information given in the problem, the relation below exists between the magnetic flux linkage and the current of the magnetic core:

$$\lambda = 0.3\sqrt{i} \tag{1}$$

Moreover, we have:

$$i = 4 \, A \tag{2}$$

From Eq. (1), we have:

$$d\lambda = 0.3 \times \frac{1}{2\sqrt{i}} di \tag{3}$$

The energy saved in the magnetic circuit can be calculated as follows:

$$W_f = \int i d\lambda \tag{4}$$

Solving (3) and (4):

$$\Rightarrow W_f = \int \frac{0.3}{2} \sqrt{i} di \tag{5}$$

The energy saved in the magnetic circuit for $i = 4 \, A$ can be calculated as follows:

$$\Rightarrow W_f = \int_0^4 \frac{0.3}{2} \sqrt{i} di = 0.1 i^{\frac{3}{2}} \Big|_0^4 = 0.8 \, J$$

Choice (4) is the answer.

8.8. Based on the information given in the problem, the relation below exists between the magnetic flux linkage and the current of the magnetic circuit of an electromechanical relay:

$$\lambda = \frac{4}{(x+1)} \left[i^{\frac{1}{2}} + i^{\frac{1}{3}} \right] \tag{1}$$

Moreover, we have:

$$i = 1 \, A \tag{2}$$

$$x = 1 \, m \tag{3}$$

Since the relation between the magnetic flux linkage and the current has been given in the form of $\lambda = \lambda(i, x)$, the current is an independent variable, and we can determine the co-energy saved in the magnetic circuit as follows:

$$W_f' = \int \lambda di \tag{4}$$

Solving (1) and (4):

$$W_f' = \int \frac{4}{x+1} \left(i^{\frac{1}{2}} + i^{\frac{1}{3}} \right) di = \frac{4}{x+1} \left(\frac{2}{3} i^{\frac{3}{2}} + \frac{3}{4} i^{\frac{4}{3}} \right) \tag{5}$$

The average electromagnetic force exerted on the moving part of the relay can be calculated as follows:

$$F_{ave} = \frac{\partial W_f'}{\partial x} \tag{6}$$

Solving (5) and (6):

$$F_{ave} = -\frac{4}{(x+1)^2} \left[\frac{2}{3} i^{\frac{3}{2}} + \frac{3}{4} i^{\frac{4}{3}} \right]$$

The average electromagnetic force for $x = 1 \, m$ and $i = 1 \, A$ is as follows:

$$\Rightarrow F_{ave} = -\frac{17}{12} \, N$$

The negative value of the average force implies that the magnetic system intends to decrease the air gap. In other words, the force is attractive.

Choice (1) is the answer.

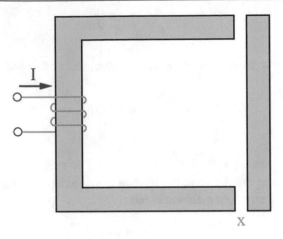

Fig. 8.3 The magnetic circuit of solution of problem 8.8

8.9. Based on the information given in the problem, we have:

$$i_s(t) = I_m \cos \omega t \tag{1}$$

$$\mu_r \to \infty \tag{2}$$

The reluctance of magnetic circuit only includes the air gap reluctance which is as follows:

$$R_g = \frac{g}{\mu_0 A} \tag{3}$$

The inductance of coil can be calculated as follows:

$$L = \frac{N^2}{R_g} = \frac{N^2 \mu_0 A}{g} \tag{4}$$

The average electromagnetic force exerted on the moving part can be calculated as follows:

$$F_{ave} = \frac{1}{2} I_{rms}^2 \frac{dL}{dg} \tag{5}$$

Solving (1), (4), and (5):

$$F_{ave} = \frac{1}{2} \left(\frac{I_m}{\sqrt{2}} \right)^2 \frac{d}{dg} \left(\frac{N^2 \mu_0 A}{g} \right) = -\frac{1}{4} \frac{N^2 I_m^2 \mu_0 A}{g^2}$$

The negative value of the average force implies that the magnetic system intends to decrease the air gap. In other words, the force is attractive.

Choice (4) is the answer.

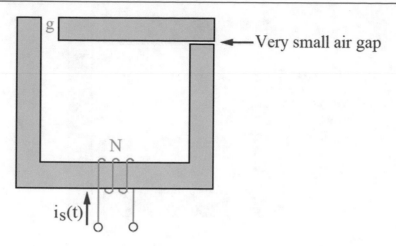

Fig. 8.4 The magnetic circuit of solution of problem 8.9

8.10. Based on the information given in the problem, we have:

$$i_s(t) = I_0 + I_0 \sin \omega t \tag{1}$$

$$\mu_r \to \infty \tag{2}$$

The reluctance of magnetic circuit only includes the equivalent reluctance of the two air gaps which is as follows:

$$R_{eq} = 2R_x = \frac{2x}{\mu_0 A} \tag{3}$$

The inductance of coil can be calculated as follows:

$$L = \frac{N^2}{R_x} = \frac{N^2 \mu_0 A}{2x} \tag{4}$$

The current includes two parts, one part is sinusoidal (ac), that is, $I_0 \sin \omega t$, and the other one is constant (DC), that is I_0. Since the system is linear, we can use superposition theorem to calculate the average electromagnetic force as follows:

$$F_{ave} = \frac{1}{2} I_0^2 \frac{dL}{dx} + \frac{1}{2} I_{rms}^2 \frac{dL}{dx} \tag{5}$$

Solving (1), (4), and (5):

$$F_{ave} = \frac{1}{2} \left(I_0^2 + \left(\frac{I_0}{\sqrt{2}} \right)^2 \right) \frac{d}{dx} \left(\frac{N^2 \mu_0 A}{2x} \right) = -\frac{3}{8} \frac{N^2 I_0^2 \mu_0 A}{x^2}$$

The negative value of the average force implies that the magnetic system intends to decrease the air gap. In other words, the force is attractive.

Choice (4) is the answer.

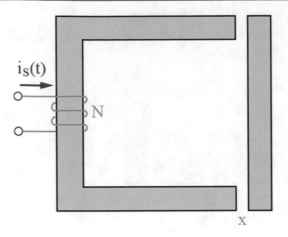

Fig. 8.5 The magnetic circuit of solution of problem 8.10

8.11. Based on the information given in the problem, we have:

$$|F_{ave}|_{g=1\ cm} = 11.25\ N \tag{1}$$

$$\mu_c \to \infty \tag{2}$$

$$\mu_0 = 10^{-6}\ H/m \tag{3}$$

$$A = 20\ cm^2 \tag{4}$$

$$N = 1000 \tag{5}$$

Since wood is not a magnetic material (its relative permeability is almost unit), the reluctance of magnetic circuit only includes the equivalent reluctance of the two air gaps which is as follows:

$$R_{eq} = 2R_g = \frac{2g}{\mu_0 A} \tag{6}$$

The inductance of coil can be calculated as follows:

$$L = \frac{N^2}{R_g} = \frac{N^2 \mu_0 A}{2g} \tag{7}$$

The average electromagnetic force exerted on the moving part can be calculated as follows:

$$F_{ave} = \frac{1}{2} I^2 \frac{dL}{dg} \tag{8}$$

Solving (7) and (8):

$$|F_{ave}| = \left| \frac{1}{2} I^2 \frac{d}{dg} \left(\frac{N^2 \mu_0 A}{2g} \right) \right| = \frac{1}{4} \frac{N^2 I^2 \mu_0 A}{g^2} \tag{9}$$

Solving (9) by considering the value of parameters:

$$11.25 = \frac{1}{4} \frac{1000^2 \times I^2 \times 10^{-6} \times 20 \times 10^{-4}}{\left(10^{-2}\right)^2}$$

$$\Rightarrow I = 1.5 \, A$$

Choice (1) is the answer.

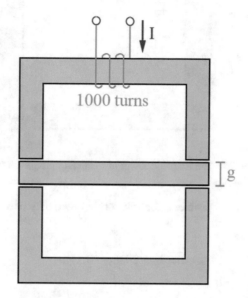

Fig. 8.6 The magnetic circuit of solution of problem 8.11

8.12. Based on the information given in the problem, the magnetic core is ideal; therefore, $\mu_c \to \infty$. Moreover, we have:

$$\varphi = 1 \, mWb \tag{1}$$

$$x = 10 \, mm \tag{2}$$

$$A = 25 \, cm^2 \tag{3}$$

Since, the core is ideal, the whole magnetic energy is saved in the air gaps and can be calculated as follows:

$$W_f = \frac{1}{2} R_{eq} \varphi^2 \tag{4}$$

where the equivalent reluctance is as follows:

$$R_{eq} = 2R_x = \frac{2x}{\mu_0 A} \tag{5}$$

Solving (1)–(5):

$$W_f = \frac{1}{2} \left(\frac{2x}{\mu_0 A}\right) \varphi^2 = \frac{1}{2} \times \frac{2 \times 10 \times 10^{-3}}{4\pi \times 10^{-7} \times 25 \times 10^{-4}} \times \left(10^{-3}\right)^2$$

$$\Rightarrow W_f = 3.18\ J$$

Choice (2) is the answer.

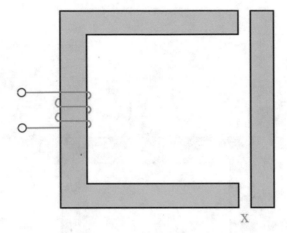

Fig. 8.7 The magnetic circuit of solution of problem 8.12

8.13. Based on the information given in the problem, the relation below exists between the magnetic flux linkage and the current of the magnetic circuit:

$$i = \lambda^{\frac{1}{2}} + 20\lambda(x - 0.1)^2 \tag{1}$$

In addition, we have:

$$x = 0.05\ m \tag{2}$$

Since the relation between the magnetic flux linkage and the current has been given in the form of $i = i(\lambda, x)$, the flux linkage is an independent variable, and we can determine the energy saved in the magnetic circuit as follows:

$$W_f = \int i\, d\lambda \tag{3}$$

Solving (1) and (3):

$$W_f = \int_0^\lambda \left[\lambda^{\frac{1}{2}} + 20\lambda(x - 0.1)^2\right] d\lambda = \frac{2}{3}\lambda^{\frac{3}{2}} + 10\lambda^2(x - 0.1)^2 \tag{4}$$

The average electromagnetic force exerted on the moving part of the relay can be calculated as follows:

$$F_{ave} = -\frac{\partial W_f}{\partial x} \tag{5}$$

Solving (4) and (5):

$$F_{ave} = -20\lambda^2(x - 0.1)$$

$$\xrightarrow{x\ =\ 0.05\ m} F_{ave} = \lambda^2$$

Choice (1) is the answer.

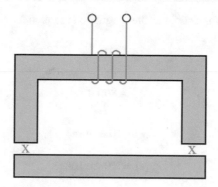

Fig. 8.8   The magnetic circuit of solution of problem 8.13

8.14. Based on the information given in the problem, the relation below exists between the magnetic flux linkage and the current of the magnetic circuit:

$$\lambda = 0.66 i^{\frac{1}{3}} \tag{1}$$

$$\Rightarrow i = \frac{\lambda^3}{(0.66)^3} \tag{2}$$

Moreover, we know that:

$$i = 1\,A \tag{3}$$

Solving (1) and (3):

$$\Rightarrow \lambda = 0.66\,Wb \tag{4}$$

Method 1: Since the relation between the magnetic flux linkage and the current in Eq. (2) is now in the form of $i = i(\lambda)$, the flux linkage is an independent variable, and we can determine the energy saved in the magnetic circuit as follows:

$$W_f = \int i\,d\lambda \tag{5}$$

Solving (2), (4), and (5):

$$\Rightarrow W_f = \int_0^{0.66} \frac{\lambda^3}{(0.66)^3}\,d\lambda \Rightarrow W_f = 0.165\,J$$

Method 2: From (1), we have:

$$\overset{d}{\Longrightarrow} d\lambda = 0.22 i^{-\frac{2}{3}}\,di \tag{6}$$

Solving (3), (5), and (6):

$$W_f = \int_0^1 i\left(0.22 i^{-\frac{2}{3}}\,di\right) = \int_0^1 0.22 i^{\frac{1}{3}}\,di = \frac{0.22}{\frac{4}{3}} i^{\frac{4}{3}}\Big|_0^1$$

$$\Rightarrow W_f = 0.165\,J$$

Choice (1) is the answer.

8.15. Based on the information given in the problem, the magnetic core is ideal. Moreover, we have:

$$A = 10 \ cm^2 \tag{1}$$

$$\frac{1}{\mu_0} \triangleq 800{,}000 \tag{2}$$

$$N = 1000 \ \text{turns} \tag{3}$$

$$v(t) = 20 \sin 100t \tag{4}$$

In an ac system, the average electromagnetic force exerted on the moving part can be calculated as follows:

$$F_{ave} = -\frac{1}{2} \varphi_{rms}^2 \frac{dR_{eq}}{dg} \tag{5}$$

where the equivalent reluctance can be determined as follows:

$$R_{eq} = 2R_g = \frac{2g}{\mu_0 A} \tag{6}$$

Solving (5) and (6):

$$F_{ave} = -\frac{1}{2} \varphi_{rms}^2 \frac{d}{dg}\left(\frac{2g}{\mu_0 A}\right) = -\frac{\varphi_{rms}^2}{\mu_0 A} \tag{7}$$

Based on Faraday's law, the relation below exists:

$$V_{max} \approx E_{max} = N\varphi_{max}\omega \tag{8}$$

$$\Rightarrow \varphi_{max} = \frac{V_{max}}{N\omega} = \frac{20}{1000 \times 100} = 2 \times 10^{-4} \ Wb \tag{9}$$

$$\Rightarrow \varphi_{rms} = \frac{\varphi_{max}}{\sqrt{2}} = \sqrt{2} \times 10^{-4} \ Wb \tag{10}$$

Solving (7) by considering the value of parameters:

$$|F_{ave}| = \left|-\frac{\left(\sqrt{2} \times 10^{-4}\right)^2}{\frac{1}{800000} \times 10 \times 10^{-4}}\right| \Rightarrow |F_{ave}| = 16 \ N$$

Choice (2) is the answer.

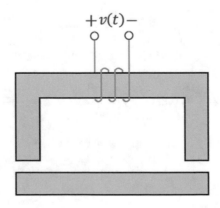

Fig. 8.9 The magnetic circuit of solution of problem 8.15

8.16. Based on the information given in the problem, the relation below exists between the magnetic flux linkage and the current of the magnetic circuit:

$$\lambda = \sqrt{\frac{i}{x}} \tag{1}$$

Moreover, we have:

$$i = 1 \, A \tag{2}$$

$$x = 1 \, m \tag{3}$$

We see that the relation between the magnetic flux linkage and the current is in the form of $\lambda = \lambda(i, x)$; therefore, the current is an independent variable, and we can determine the co-energy saved in the magnetic circuit as follows:

$$W'_f = \int \lambda di \tag{4}$$

Solving (1) and (4):

$$W'_f = \frac{1}{\sqrt{x}} \int i^{\frac{1}{2}} di = \frac{1}{\sqrt{x}} \times \frac{2}{3} i^{\frac{3}{2}} \tag{5}$$

The average electromagnetic force exerted on the moving part can be calculated as follows:

$$F_{ave} = \frac{\partial W'_f}{\partial x} \tag{6}$$

Solving (5) and (6):

$$F_{ave} = -\frac{1}{2x\sqrt{x}} \times \frac{2}{3} \times i^{\frac{3}{2}} \tag{7}$$

Solving (7) for $i = 1 \, A$ and $x = 1 \, m$:

$$F_{ave} = -\frac{1}{3} N$$

Choice (4) is the answer.

The negative value of the average force implies that the magnetic system intends to decrease the air gap. In other words, the force is attractive.

8.17. Based on the information given in the problem, we have:

$$\mu_c \rightarrow \infty \tag{1}$$

$$A = 10 \, cm^2 \tag{2}$$

$$N = 1000 \tag{3}$$

$$i = 10 \, A \tag{4}$$

$$g = 5 \, mm \tag{5}$$

The equivalent electrical circuit of the magnetic circuit is illustrated in Fig. 8.10.b. The equivalent reluctance of the magnetic circuit can be calculated as follows:

$$R_{eq} = 2R_g = \frac{2g}{\mu_0 A} \tag{6}$$

The inductance of coil can be calculated as follows:

$$L = \frac{N^2}{R_{eq}} = \frac{N^2 \mu_0 A}{2g} \tag{7}$$

The average electromagnetic force exerted on the moving part can be calculated as follows:

$$F_{ave} = \frac{1}{2} I^2 \frac{dL}{dg} = -\frac{1}{2} I^2 \frac{N^2 \mu_0 A}{2g^2} \tag{8}$$

Solving (8) by considering the value of parameters:

$$\Rightarrow F_{ave} = -\frac{1}{2} \times 10^2 \times \frac{1000^2 \times 4\pi \times 10^{-7} \times 10 \times 10^{-4}}{2 \times (5 \times 10^{-3})^2}$$

$$\Rightarrow |F_{ave}| = 4\pi \, N$$

Choice (3) is the answer.

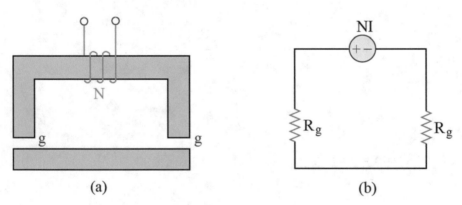

Fig. 8.10 The magnetic circuit and equivalent electrical circuit of solution of problem 8.17

8.18. Based on the information given in the problem, we have:

$$L_{sr}(\theta) = 0.04 \cos \theta \tag{1}$$

$$i_s = i_r = 10\sqrt{2} \sin \omega t \tag{2}$$

The electromagnetic torque exerted on the rotor can be calculated as follows:

$$T = i_{s,rms} i_{r,rms} \frac{dL_{sr}(\theta)}{d\theta} \tag{3}$$

Solving (1), (2) and (3):

$$T = (10 \sin \omega t)(10 \sin \omega t) \frac{d}{d\theta}(0.04 \cos \theta) = -100 \sin^2 \omega t \times 0.04 \sin \theta \, N.m$$

$$T = -100\left(\frac{1 - \cos 2\omega t}{2}\right)(0.04 \sin \theta) = -2 \sin \theta + 2 \sin \theta \cos 2\omega t \qquad (4)$$

The average electromagnetic torque is as follows:

$$T_{ave} = -2 \sin \theta \; N.m \qquad (5)$$

In addition, the maximum value of the average electromagnetic torque is as follows:

$$\Rightarrow |T_{ave, max}| = 2 \; N.m$$

Choice (2) is the answer.

8.19. Based on the information given in the problem, the energy saved in the magnetic circuit of the electric machine is as follows:

$$W = 5i_s^2 - 2i_r i_s \cos \theta + 0.04 i_r^2 \cos 2\theta \qquad (1)$$

As we know, the general equation of energy saved in a rotational two-part electromechanical system is as follows:

$$W_f(i_s, i_r, \theta) = \frac{1}{2} L_{ss}(\theta)i_s^2 + \frac{1}{2} L_{rr}(\theta)i_r^2 + L_{sr}(\theta)i_s i_r \qquad (2)$$

By comparing (1) and (2), we have the following results:

$$L_{ss}(\theta) = \frac{5}{\frac{1}{2}} = 10 \; H \qquad (3)$$

$$L_{rr}(\theta) = \frac{0.04}{\frac{1}{2}} \cos 2\theta = 0.08 \cos 2\theta \; H \qquad (4)$$

$$L_{sr}(\theta) = -2 \cos \theta \; H \qquad (5)$$

As can be noticed from (4), L_{rr} is a function of θ; thus, the stator has salient poles. However, as can be seen in (3), L_{ss} does not depend on θ; therefore, the rotor is cylindrical. Choice (1) is the answer.

8.20. Based on the information given in the problem, we have:

$$L_{11} = 0.1 \; H \qquad (1)$$

$$L_{22} = 0.15 \; H \qquad (2)$$

$$L_{12} = 0.04 \cos \theta \; H \qquad (3)$$

$$i_1(t) = i_2(t) = \sin \omega t \; A \qquad (4)$$

Since the self-inductances are independent of the rotation angle, only electromagnetic torque will be developed. In other words, there will be no reluctance torque.

The electromagnetic torque exerted on the moving part can be calculated as follows:

$$T = i_{1,rms} i_{2,rms} \frac{dL_{12}(\theta)}{d\theta} \qquad (5)$$

Solving (3), (4), and (5):

$$T = \sin^2 \omega t \frac{d}{d\theta}(0.04 \cos \theta) = -0.04 \sin \theta \sin^2 \omega t \ N.m$$

$$T = -0.04 \sin \theta \left(\frac{1}{2} - \frac{1}{2} \cos 2\omega t\right) = -0.02 \sin \theta + 0.02 \sin \theta \cos 2\omega t \ N.m$$

The average electromagnetic torque is as follows:

$$T_{ave} = -0.02 \sin \theta \ N.m$$

Moreover, the maximum value of the average electromagnetic torque is as follows:

$$\Rightarrow |T_{ave, max}| = 0.02 \ N.m$$

Choice (2) is the answer.

8.21. Based on the information given in the problem, the inductance matrix of the electric machine is as follows:

$$L = \begin{bmatrix} 1 & 0.5 \cos \theta \\ 0.5 \cos \theta & 1.5 \end{bmatrix} H \tag{1}$$

$$i_s(t) = 5 \ A \tag{2}$$

$$i_r(t) = 10 \ A \tag{3}$$

$$\theta = 90° \tag{4}$$

The electromagnetic torque exerted on the rotor can be calculated as follows:

$$T = \frac{1}{2} i_s^2 \frac{dL_{ss}(\theta)}{d\theta} + \frac{1}{2} i_r^2 \frac{dL_{rr}(\theta)}{d\theta} + i_s i_r \frac{dL_{sr}(\theta)}{d\theta} \tag{5}$$

Solving (1)–(3) and (5):

$$T = 0 + 0 + 5 \times 10 \times (-0.5 \sin \theta) = -25 \sin \theta \tag{6}$$

Solving (4) and (6):

$$\Rightarrow T_{ave} = -25 \ N.m$$

Choice (2) is the answer.

The negative value of the average torque implies that the magnetic system intends to decrease the air gap. In other words, the torque is attractive.

8.22. Based on the information given in the problem, the developed torque of the electric machine is as follows:

$$T = 2 i_s i_r \sin \theta - 0.04 i_r^2 \sin 2\theta \tag{1}$$

As we know, the general equation of torque developed in an electric machine is as follows:

$$T = \frac{1}{2} i_s^2 \frac{dL_{ss}(\theta)}{d\theta} + \frac{1}{2} i_r^2 \frac{dL_{rr}(\theta)}{d\theta} + i_s i_r \frac{dL_{sr}(\theta)}{d\theta} \tag{2}$$

By comparing (1) and (2), we have the following results:

$$\frac{dL_{ss}(\theta)}{d\theta} = 0 \tag{3}$$

$$\frac{dL_{rr}(\theta)}{d\theta} \neq 0 \tag{4}$$

Therefore, the stator has salient poles, but the rotor is cylindrical. Choice (4) is the answer.

8.23. Based on the information given in the problem, the inductance matrix of the electromagnetic relay is as follows:

$$L = \begin{bmatrix} L_{11} & L_{12} \\ L_{21} & L_{22} \end{bmatrix} = \begin{bmatrix} 1+x & 1-x \\ 1-x & 2+x \end{bmatrix} \tag{1}$$

Moreover, we have:

$$|I_1| = 1 \, A \tag{2}$$

$$|I_2| = 2 \, A \tag{3}$$

The electromagnetic force exerted on the moving part can be calculated as follows:

$$F = \frac{1}{2}I_1^2\frac{dL_{11}}{dx} + I_1 I_2 \frac{dL_{12}}{dx} + \frac{1}{2}I_2^2\frac{dL_{22}}{dx} \tag{4}$$

$$\Rightarrow F = \frac{1}{2}I_1^2 - I_1 I_2 + \frac{1}{2}I_2^2 \tag{5}$$

Since the repulsive electromagnetic force has been requested, the currents must have different signs. Therefore:

$$F = \left(\frac{1}{2} \times 1^2\right) - \left(1 \times (-2)\right) + \left(\frac{1}{2} \times 2^2\right)$$

$$\Rightarrow F = 4.5 \, N$$

Choice (2) is the answer.

8.24. The equivalent electrical circuit of the magnetic circuit is illustrated in Fig. 8.11.b. The inductance of coil can be calculated as follows:

$$L = \frac{N^2}{R_{eq}} \tag{1}$$

The equivalent reluctance can be determined as follows:

$$R_{eq} = R_x + R_x \| R_x = \frac{3}{2}R_x \tag{2}$$

where:

$$R_x = \frac{x}{\mu_0 A} \tag{3}$$

Solving (1)–(3):

$$L = \frac{N^2}{\frac{3}{2}\frac{x}{\mu_0 A}} = \frac{2}{3}\frac{\mu_0 A N^2}{x} \tag{4}$$

Since the air gap decreases from x_1 to x_2 with an infinite speed, the flux linkage will remain constant:

$$\lambda_1 = \lambda_2 \Rightarrow L_1 I_1 = L_2 I_2 \tag{5}$$

Solving (4) and (5):

$$\frac{2}{3}\frac{\mu_0 A N^2}{x_1} I_1 = \frac{2}{3}\frac{\mu_0 A N^2}{x_2} I_2 \tag{6}$$

$$\Rightarrow \frac{I_1}{I_2} = \frac{x_1}{x_2}$$

Choice (1) is the answer.

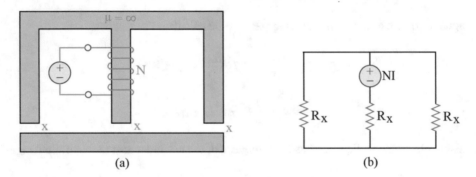

Fig. 8.11 The magnetic circuit and equivalent electrical circuit of solution of problem 8.24

8.25. The average electromagnetic torque exerted on the moving part can be calculated as follows:

$$T_{ave} = \frac{1}{2} I^2 \frac{dL}{d\theta} \tag{1}$$

where the inductance of coil can be calculated as follows:

$$L = \frac{N^2}{R} \tag{2}$$

The reluctance of magnetic circuit only includes the air gap reluctance which is as follows:

$$R = \frac{r\theta}{\mu_0 A} \tag{3}$$

Solving (1)–(3):

$$\Rightarrow T_{ave} = \frac{1}{2} \times I^2 \frac{d}{d\theta}\left(\frac{\mu_0 A N^2}{r\theta}\right) = \frac{1}{2} I^2 \left(-\frac{\mu_0 A N^2}{r\theta^2}\right) \tag{4}$$

Let us assume:

$$k \triangleq \frac{1}{2r} \mu_0 A N^2 \tag{5}$$

Solving (4) and (5):

$$\Rightarrow T_{ave} = -k \left(\frac{I}{\theta} \right)^2$$

Choice (1) is the answer.

The negative value of the average torque implies that the magnetic system intends to decrease the air gap. In other words, the torque is attractive.

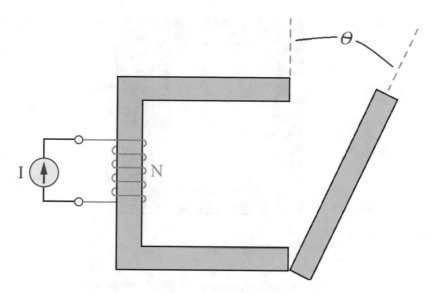

Fig. 8.12 The magnetic circuit of solution of problem 8.25

8.26. Based on the information given in the problem, the magnetic core is ideal. Moreover, we have:

$$B_y = 0.4\ T \tag{1}$$

$$A_A = 2\ cm^2 \tag{2}$$

$$A_B = 4\ cm^2 \tag{3}$$

The system includes only one loop; therefore, the magnetic flux will be the same throughout the core. In other words:

$$\varphi_x = \varphi_y = B_y A_B = 0.4 \times 4 \times 10^{-4} = 1.6 \times 10^{-4} Wb \tag{4}$$

The reluctance of the air gaps of g_x and g_y can be calculated as follows:

$$R_x = \frac{g_x}{\mu_0 A_A} = \frac{g_x}{4\pi \times 10^{-7} \times 2 \times 10^{-4}} = \frac{g_x}{8\pi \times 10^{-11}}\ \frac{A}{Wb} \tag{5}$$

$$R_y = \frac{g_y}{\mu_0 A_B} = \frac{g_y}{4\pi \times 10^{-7} \times 4 \times 10^{-4}} = \frac{g_y}{16\pi \times 10^{-11}}\ \frac{A}{Wb} \tag{6}$$

The average electromagnetic force exerted on the moving part in both directions can be calculated as follows:

$$F_x = -\frac{1}{2}\varphi_x^2 \frac{dR_x}{dg_x} = -\frac{1}{2} \times \left(1.6 \times 10^{-4}\right)^2 \times \frac{1}{8\pi \times 10^{-11}} \Rightarrow F_x = -\frac{160}{\pi}N$$

$$F_y = -\frac{1}{2}\varphi_y^2 \frac{dR_y}{dg_y} = -\frac{1}{2} \times \left(1.6 \times 10^{-4}\right)^2 \times \frac{1}{16\pi \times 10^{-11}} \Rightarrow F_y = -\frac{80}{\pi}N$$

Choice (4) is the answer.

The negative value of the average forces implies that the magnetic system intends to decrease the air gaps. In other words, both forces are attractive.

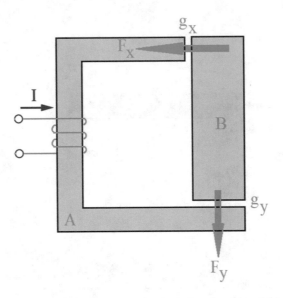

Fig. 8.13   The magnetic circuit of solution of problem 8.26

8.27. Based on the information given in the problem, a copper plate with the thickness of *d mm* has been surrounded by the U-shape magnets with the cross-sectional area of *A cm²* and the relative permeability of $\mu_r \gg 1$. Moreover, there is no flux leakage in the system.

Since copper is not a magnetic material (its relative permeability is almost unit), the reluctance of the magnetic circuit only includes the equivalent reluctance of the two copper gaps which is as follows:

$$R_{eq} = 2R_{cu} = \frac{2d}{\mu_0 A}$$

Figure 8.14 shows the equivalent electrical circuit of the magnetic system. It should be noted that the coils have opposing magnetic fluxes due to their winding directions. The equivalent inductance of the coils can be calculated as follows:

$$L = \frac{N_{eq}^2}{R_{eq}} = \frac{(N_1 - N_2)^2}{\frac{2d}{\mu_0 A}} = \frac{\mu_0 A (N_1 - N_2)^2}{2d}$$

The absolute value of force exerted on each side of the copper plate can be calculated as follows:

$$F_{ave} = \frac{1}{2}I^2 \frac{dL}{dd} = -\frac{1}{2}I^2 \frac{\mu_0 A (N_1 - N_2)^2}{2d^2} \tag{3}$$

$$\Rightarrow |F_{ave}| = \mu_0 A \frac{(N_1 - N_2)^2 I^2}{4d^2}$$

Choice (1) is the answer.

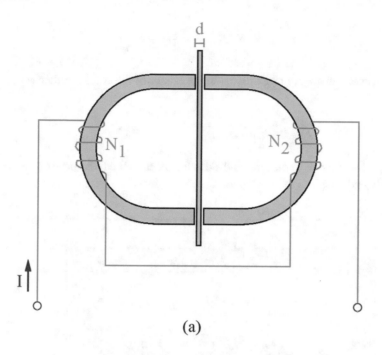

(a)

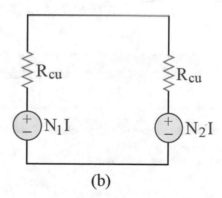

(b)

Fig. 8.14 The magnetic circuit and equivalent electrical circuit of solution of problem 8.27

8.28. The average electromagnetic force exerted on the moving part can be calculated as follows:

$$F_{ave} = -\frac{1}{2} \varphi^2 \frac{dR}{dg} \tag{1}$$

The reluctance of the air gaps can be calculated as follows:

$$R_1 = \frac{g_0}{\mu_0 h_{right} d}$$

$$R_2 = \frac{g_0}{\mu_0 h_{left} d}$$

where h and d are the height and the thickness of the core, respectively.

The average electromagnetic force exerted on the right-hand side and left-hand side of the moving part can be calculated as follows:

$$\Rightarrow F_{ave,right} = -\frac{1}{2}\varphi^2 \frac{1}{\mu_0 dh_{right}}$$

$$\Rightarrow F_{ave,left} = -\frac{1}{2}\varphi^2 \frac{1}{\mu_0 dh_{left}}$$

As can be seen, the force is inversely proportional to the height of the core. In other words:

$$\frac{F_{ave,right}}{F_{ave,left}} = \frac{h_{left}}{h_{right}} \xrightarrow{h_{right} > h_{left}} F_{ave,right} < F_{ave,left}$$

Hence, the moving part will move to the left-hand side and attach to the fixed part. Or, we can say that the moving part will move to one side and attach to the fixed part. Choice (4) is the answer.

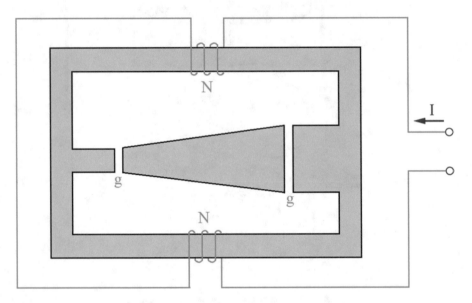

Fig. 8.15 The magnetic circuit of solution of problem 8.28

8.29. Based on the information given in the problem, we have:

$$\mu_c \to \infty \tag{1}$$

$$g = 1 \ mm \tag{2}$$

$$d = 10 \ cm \tag{3}$$

$$a = 2.5 \ cm \tag{4}$$

$$N = 200 \tag{5}$$

$$I_{rms} = 1 \ A \tag{6}$$

The equivalent electrical circuit of the magnetic circuit is illustrated in Fig. 8.16.b. The equivalent reluctance of the magnetic circuit can be calculated as follows:

$$R_{eq} = R_2 + (R_1 \parallel R_1) \tag{7}$$

where:

$$R_1 = \frac{g}{\mu_0 da} \tag{8}$$

$$R_2 = \frac{g}{\mu_0 d(2a)} \tag{9}$$

Therefore:

$$R_{eq} = \frac{g}{2\mu_0 da} + \frac{g}{2\mu_0 da} = \frac{g}{\mu_0 da} \tag{10}$$

The inductance of the coil can be calculated as follows:

$$L = \frac{N^2}{R_{eq}} = \frac{N^2 \mu_0 da}{g} \tag{11}$$

The average electromagnetic force exerted on the moving part can be calculated as follows:

$$F_{ave} = \frac{1}{2}I_{rms}^2 \frac{dL}{dg} \tag{12}$$

Solving (11) and (12):

$$F_{ave} = \frac{1}{2}I_{rms}^2 \times \frac{-N^2 \mu_0 da}{g^2} \tag{13}$$

Solving (13) and considering the value of parameters:

$$F_{ave} = -\frac{1}{2} \times 1^2 \times \frac{200^2 \times 4\pi \times 10^{-7} \times 0.1 \times 0.025}{0.001^2}$$

$$\Rightarrow F_{ave} = -20\pi \, N$$

Choice (1) is the answer.

The negative value of the average force implies that the magnetic system intends to decrease the air gap. In other words, the force is attractive.

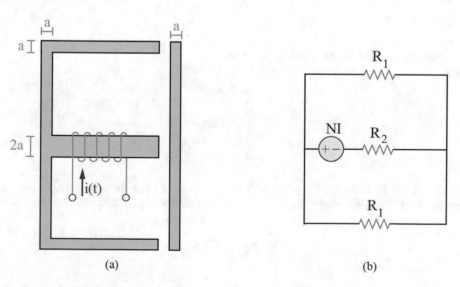

Fig. 8.16 The magnetic circuit and equivalent electrical circuit of solution of problem 8.29

8.30. Based on the information given in the problem, we have:

$$W_f(\lambda, x) = \frac{\lambda^3}{0.1 - x} \tag{1}$$

$$i = 3 \, A \tag{2}$$

$$x = 0.02 \, m \tag{3}$$

Based on the equation of energy saved in the magnetic field, we have:

$$\partial W_f = id\lambda \Rightarrow i = \frac{\partial W_f(\lambda, x)}{\partial \lambda} \tag{4}$$

Solving (1) and (4):

$$i = \frac{3\lambda^2}{0.1 - x} \tag{5}$$

Solving (2), (3), and (5):

$$\lambda = 0.2\sqrt{2} \, Wb \tag{6}$$

The average electromagnetic force exerted on the moving part can be calculated as follows:

$$F_{ave} = -\frac{\partial W_f(\lambda, x)}{\partial x} \tag{7}$$

Solving (7) by using the value of parameters:

$$F_{ave} = -\frac{\partial}{\partial x}\left(\frac{\lambda^3}{0.1 - x}\right) = -\frac{\lambda^3}{(0.1 - x)^2}$$

$$\Rightarrow F_{ave} = -\frac{\left(0.2\sqrt{2}\right)^3}{(0.1 - 0.02)^2} \Rightarrow F_{ave} = -3.53 \, N$$

Choice (4) is the answer.

8.31. Based on the information given in the problem, we have:

$$A = 16 \, cm^2 \tag{1}$$

$$N = 210 \text{ turns} \tag{2}$$

$$I = 2 \, A \tag{3}$$

As we know, by writing Kirchhoff's Magnetomotive Force Law (KML) for a magnetic circuit and considering the direction of magnetic flux of coils, we have:

$$\sum_{i=1}^{m} F_{m,i} = \sum_{j=1}^{n} V_{m,j} \tag{4}$$

where:

$$\sum_{i=1}^{m} F_{m,i} = \sum_{i=1}^{m} N_i I_i \tag{5}$$

$$\sum_{j=1}^{n} V_{m,j} = \sum_{j=1}^{n} R_j \varphi_j = \sum_{j=1}^{n} H_j l_j \tag{6}$$

Equations (4)–(6) state that sum of magnetomotive forces (mmf) in a loop ($\sum_{i=1}^{m} F_{m,i}$), achieved from the ampere-turn excitation of the coils, is equal to sum of mmf drops ($\sum_{j=1}^{n} V_{m,j}$), achieved from the product of flux and reluctances or the product of magnetic field intensity and length of magnetic components, across the rest of the loop.

Now, by applying (4)–(6) on the magnetic circuit of Fig. 8.17.a, we can write:

$$NI = H l_{ave} \tag{7}$$

$$\Rightarrow H = \frac{NI}{l_{ave}} = \frac{2 \times 210}{2 \times (15 + 20) \times 10^{-2}} = 600 \frac{A}{m} \tag{8}$$

From the $B - H$ curve of the core, for $H = 600 \frac{A}{m}$, we have:

$$B = 1.5\ T \tag{9}$$

The magnetic energy saved in the core can be calculated as follows:

$$W_f = (\text{Magnetic energy density}) \times (\text{Volume of core}) = w_f \times V_{core} \tag{10}$$

As we know, the area above the B-H curve is equal to the magnetic energy density. Therefore:

$$w_f = \frac{100 \times 1}{2} + \frac{(600 + 100) \times 0.5}{2} = 225 \frac{J}{m^3} \tag{11}$$

The volume of the core can be determined as follows:

$$V_{core} = l_{ave} \times A = \left(2 \times (15 + 20) \times 10^{-2}\right)\left(16 \times 10^{-4}\right) = 1.12 \times 10^{-3}\ m^3 \tag{12}$$

Solving (10)–(12):

$$\Rightarrow W_f = 225 \times \left(1.12 \times 10^{-3}\right) = 0.252\ J$$

Choice (3) is the answer.

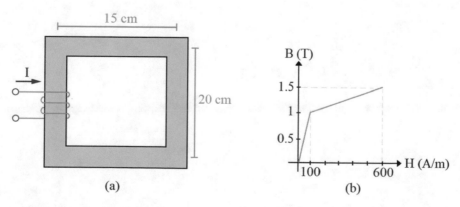

Fig. 8.17 The magnetic circuit and equivalent electrical circuit of solution of problem 8.31

Problems: Separately Excited DC Electric Generator

Abstract

In this chapter, the problems concerned with the separately excited DC electric generators are solved. In this chapter, the problems are categorized in different levels based on their difficulty levels (easy, normal, and hard) and calculation amounts (small, normal, and large). Additionally, the problems are ordered from the easiest problem with the smallest computations to the most difficult problems with the largest calculations.

9.1. In a four-pole DC generator that rotates with the speed of 750 rpm, the electromotive force (emf) of 240 is generated. If the magnetic flux under each pole is about 14.5 mWb, calculate the fringing factor. Herein, the armature includes a wave winding with 792 as the total number of conductors of armature winding.

Difficulty level ○ Easy ● Normal ○ Hard
Calculation amount ● Small ○ Normal ○ Large
1) 1.1
2) 1.05
3) 1.2
4) 1.25

9.2. In a 360 V DC machine, the armature has 36 slots where each slot includes ten conductors. If the type of winding is simple lap and the speed of the machine is 1000 rpm, how much is the magnetic flux under each pole? Herein, the armature winding resistance is ignorable.

Difficulty level ○ Easy ● Normal ○ Hard
Calculation amount ● Small ○ Normal ○ Large
1) 0.18 *Wb*
2) 0.72 *Wb*
3) 0.06 *Wb*
4) 1.44 *Wb*

9.3. A four-pole, 240 V, DC machine has 24 slots where each slot includes ten conductors. If the armature includes wave winding and the machine is rotating at 1200 rpm, calculate the magnetic flux under each pole. Assume that the armature winding resistance is negligible.

Difficulty level ○ Easy ● Normal ○ Hard
Calculation amount ● Small ○ Normal ○ Large
1) 50 *mWb*
2) 75 *mWb*
3) 25 *mWb*
4) 100 *mWb*

© The Author(s), under exclusive license to Springer Nature Switzerland AG 2022
M. Rahmani-Andebili, *DC Electric Machines, Electromechanical Energy Conversion Principles, and Magnetic Circuit Analysis*,
https://doi.org/10.1007/978-3-031-08863-6_9

9.4. A DC machine includes wave winding. If the number of poles of machine is changed, how the developed electromagnetic torque will change?

Difficulty level   ○ Easy   ● Normal   ○ Hard
Calculation amount   ● Small   ○ Normal   ○ Large

1) The electromagnetic torque will increase proportional to the number of poles.
2) The electromagnetic torque will decrease proportional to the number of poles.
3) The electromagnetic torque will remain constant.
4) The information is not enough.

9.5. A DC machine includes simple lap winding. If the number of poles of machine is changed, how the developed electromagnetic torque will change?

Difficulty level   ○ Easy   ● Normal   ○ Hard
Calculation amount   ● Small   ○ Normal   ○ Large

1) The electromagnetic torque will increase proportional to the number of poles.
2) The electromagnetic torque will decrease proportional to the number of poles.
3) The electromagnetic torque will remain constant.
4) The information is not enough.

9.6. A separately excited DC generator is connected to a 500 V power supply. Moreover, its excitation circuit is connected to another 500 V power supply. What percentage of speed of this generator needs to be reduced so that its output power decreases from 500 kW to 250 kW? Assume that the armature winding resistance is 0.015 Ω and ignore the armature reaction.

Difficulty level   ○ Easy   ● Normal   ○ Hard
Calculation amount   ○ Small   ● Normal   ○ Large

1) 1.46%
2) 2.1%
3) 2.92%
4) 4.16%

9.7. In a DC machine, if the speed and the magnetic flux of machine increase to 10% and decrease by 15%, respectively, which one of the choices below is correct?

Difficulty level   ○ Easy   ● Normal   ○ Hard
Calculation amount   ○ Small   ● Normal   ○ Large

1) The electromotive force (emf) will decrease about 3.5%.
2) The electromotive force (emf) will increase about 3.5%.
3) The electromotive force (emf) will decrease about 6.5%.
4) The electromotive force (emf) will increase about 6.5%.

9.8. A separately excited DC generator is rotating at the speed of 3000 rpm and delivers the current of 100 A to a constant resistive load at the voltage of 290 V. If the speed of generator is decreased to 2100 rpm, calculate the power delivered to the load. Assume that the armature winding resistance is about 0.1 Ω.

Difficulty level   ○ Easy   ○ Normal   ● Hard
Calculation amount   ○ Small   ● Normal   ○ Large

1) 14260 W
2) 13278 W
3) 14210 W
4) 12378 W

9.9. The no-load characteristics of a separately excited DC generator at the speed of n_0 are as follows:

E_a in Volt	20	50	100	200	300	400
I_f in Ampere	0	1	1.6	4	6	9

Calculate the total resistance of the field winding if the no-load voltage of generator at the speed of $2n_0$ is 200 V. Herein, assume that the armature winding resistance is ignorable and the excitation circuit is connected to a 200 V power supply.

Difficulty level   ○ Easy   ○ Normal   ● Hard
Calculation amount   ○ Small   ● Normal   ○ Large

1) 50 Ω
2) 62.5 Ω
3) 100 Ω
4) 125 Ω

Solutions of Problems: Separately Excited DC Electric Generator

10

Abstract

In this chapter, the problems of the ninth chapter are fully solved, in detail, step-by-step, and with different methods.

10.1. Based on the information given in the problem, the armature includes a wave winding. In addition, we have:

$$p = 4 \tag{1}$$

$$n = 750 \; rpm \tag{2}$$

$$E_{a,actual} = 240 \; V \tag{3}$$

$$\varphi = 14.5 \; mWb/pole \tag{4}$$

$$Z = 792 \tag{5}$$

As we know, electromotive force (emf) of a DC machine can be calculated as follows:

$$E_{a,theoretical} = k_a \varphi \omega = \frac{pZ}{2\pi a} \varphi \omega = \frac{pZ}{60a} \varphi n \tag{6}$$

Since the armature includes a wave winding, $a = 2$. Therefore:

$$E_a = \frac{4 \times 792}{60 \times 2} \times 14.5 \times 10^{-3} \times 750 = 287.1 \; V \tag{7}$$

$$\text{Fringing factor} = \text{Leakage factor} = \frac{E_{a,theoretical}}{E_{a,actual}} = \frac{287.1}{240} \simeq 1.2$$

Choice (3) is the answer.

10.2. Based on the information given in the problem, the winding type of armature is simple lap winding. In addition, we have:

$$V_t = 360 \tag{1}$$

$$N_s = 36 \tag{2}$$

$$N_c = 10 \tag{3}$$

© The Author(s), under exclusive license to Springer Nature Switzerland AG 2022
M. Rahmani-Andebili, *DC Electric Machines, Electromechanical Energy Conversion Principles, and Magnetic Circuit Analysis*,
https://doi.org/10.1007/978-3-031-08863-6_10

$$n = 1000 \; rpm \tag{4}$$

$$R_a \approx 0 \; \Omega \tag{5}$$

As we know, armature voltage of a DC machine can be calculated as follows:

$$E_a = k_a \varphi \omega = \frac{pZ}{2\pi a} \varphi \omega = \frac{pZ}{60a} \varphi n = \frac{p(N_s \times N_c)}{60a} \varphi n \tag{6}$$

Since the armature includes a simple lap winding, $a = p$. Moreover, since $R_a \approx 0 \; \Omega$, we have $E_a \approx V_t = 360 \; V$. Therefore:

$$360 = \frac{(36 \times 10)}{60} \times \varphi \times 1000 \Rightarrow \varphi = 0.06 \; Wb/pole$$

Choice (3) is the answer.

10.3. Based on the information given in the problem, the armature includes a wave winding. In addition, we have:

$$p = 4 \tag{1}$$

$$V_t = 240 \tag{2}$$

$$N_s = 24 \tag{3}$$

$$N_c = 10 \tag{4}$$

$$n_m = 1200 \; rpm \tag{5}$$

$$R_a \approx 0 \; \Omega \tag{6}$$

As we know, armature voltage of a DC machine can be calculated as follows:

$$E_a = k_a \varphi \omega = \frac{pZ}{2\pi a} \varphi \omega = \frac{pZ}{60a} \varphi n = \frac{p(N_s \times N_c)}{60a} \varphi n \tag{7}$$

Since the armature includes a wave winding, $a = 2$. Moreover, since $R_a \approx 0 \; \Omega$, we have $E_a \approx V_t = 240 \; V$. Therefore:

$$240 = \frac{4 \times (24 \times 10)}{60 \times 2} \times \varphi \times 1200 \Rightarrow \varphi = 25 \; mWb/pole$$

Choice (3) is the answer.

10.4. Based on the information given in the problem, the DC machine includes wave winding. Therefore:

$$a = 2 \tag{1}$$

As we know, electromagnetic torque of a DC machine can be calculated as follows:

$$T_e = k_a \varphi I_a = \frac{pZ}{2\pi a} \varphi I_a \tag{2}$$

Solving (1) and (2):

$$T_e = \frac{pZ}{4\pi}\varphi I_a \Rightarrow T_e \propto p$$

Hence, the electromagnetic torque will increase proportional to the number of poles. Choice (1) is the answer.

10.5. Based on the information given in the problem, the DC machine includes a simple lap winding. Therefore:

$$a = p \tag{1}$$

As we know, electromagnetic torque of a DC machine can be calculated as follows:

$$T_e = k_a \varphi I_a = \frac{pZ}{2\pi a}\varphi I_a \tag{2}$$

Solving (1) and (2):

$$T_e = \frac{Z}{2\pi}\varphi I_a \Rightarrow T = \text{Const.}$$

As can be seen in (3), the electromagnetic torque does not have a dependency on the number of poles of machine that includes simple lap winding; thus, it will remain constant. Choice (3) is the answer.

10.6. Based on the information given in the problem, the armature reaction is ignorable. Moreover, we have:

$$V_t = 500\ V \tag{1}$$

$$V_f = 500\ V \tag{2}$$

$$P_1 = 500\ kW \tag{3}$$

$$P_2 = 250\ kW \tag{4}$$

$$R_a = 0.015\ \Omega \tag{5}$$

Figure 10.1 shows the electrical circuit of the separately excited DC generator. The primary load current can be calculated as follows:

$$I_{t,1} = I_{a,1} = \frac{P_1}{V_t} = \frac{500 \times 10^3}{500} = 1000\ A \tag{6}$$

The corresponding induced armature voltage can be calculated by writing a KVL in the loop as follows:

$$E_{a,1} = V_t + R_a I_{a,1} = 500 + (0.015 \times 1000) = 515\ V \tag{7}$$

The secondary load current can be calculated as follows:

$$I_{t,2} = I_{a,2} = \frac{P_2}{V_t} = \frac{250 \times 10^3}{500} = 500\ A \tag{8}$$

The corresponding induced armature voltage can be calculated as follows:

$$E_{a,2} = V_t + R_a I_{a,2} = 500 + (0.015 \times 500) = 507.5\ V \tag{9}$$

As we know, the induced armature voltage can be determined by using $E_a = k_a \varphi \omega$. Therefore, we can write:

$$\frac{E_{a,1}}{E_{a,2}} = \frac{\varphi_1 \omega_1}{\varphi_2 \omega_2} \tag{10}$$

However, since the excitation circuit is connected to a constant voltage, the field current and consequently the magnetic flux will remain constant. Hence:

$$\frac{E_{a,1}}{E_{a,2}} = \frac{\omega_1}{\omega_2} \tag{11}$$

$$\Rightarrow \frac{515}{507.5} = \frac{\omega_1}{\omega_2} \Rightarrow \omega_2 = 0.9854\omega_1 \tag{12}$$

Therefore:

$$\frac{\omega_2 - \omega_1}{\omega_1} \times 100 = \frac{0.9854\omega_1 - \omega_1}{\omega_1} \times 100 = -1.46\%$$

Choice (1) is the answer.

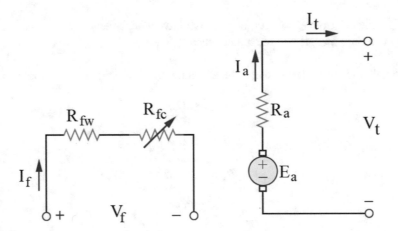

Fig. 10.1 The electrical circuit of solution of problem 10.6

10.7. Based on the information given in the problem, we have:

$$\omega_2 = 1.1\omega_1 \tag{1}$$

$$\varphi_2 = 0.85\varphi_1 \tag{2}$$

As we know, the induced armature voltage can be determined by using $E_a = k_a \varphi \omega$. Therefore, we can write:

$$\frac{E_{a,1}}{E_{a,2}} = \frac{\varphi_1 \omega_1}{\varphi_2 \omega_2} \tag{3}$$

Solving (1)–(3):

$$\frac{E_{a,1}}{E_{a,2}} = \frac{\varphi_1 \omega_1}{0.85\varphi_1 \times 1.1\omega_1} \Rightarrow E_{a,2} = 0.935E_{a,1} \tag{4}$$

Therefore:

$$\frac{E_{a,2} - E_{a,1}}{E_{a,1}} \times 100 = \frac{0.935E_{a,1} - E_{a,1}}{E_{a,1}} \times 100 = -6.5\%$$

Choice (3) is the answer.

10.8. Based on the information given in the problem, we have:

$$n_1 = 3000 \; rpm \tag{1}$$

$$I_{t,1} = 100 \; A \tag{2}$$

$$V_t = 290 \; V \tag{3}$$

$$n_2 = 2100 \; rpm \tag{4}$$

$$R_a = 0.1 \; \Omega \tag{5}$$

Figure 10.2 shows the electrical circuit of the separately excited DC generator. The load resistance can be calculated by using Ohm's law as follows:

$$R_L = \frac{V_t}{I_{t,1}} = \frac{290}{100} = 2.9 \; \Omega \tag{6}$$

Moreover, the induced voltage of armature can be calculated by applying a KVL in the right-hand side mesh as follows:

$$E_{a,1} = V_t + R_a I_{a,1} = 290 + 0.1 \times 100 = 300 \; V \tag{7}$$

As we know, the induced armature voltage can be determined by using $E_a = k_a \varphi \omega$. Therefore, we can write:

$$\frac{E_{a,1}}{E_{a,2}} = \frac{\varphi_1 \omega_1}{\varphi_2 \omega_2} = \frac{\varphi_1 n_1}{\varphi_2 n_2} \tag{8}$$

Since the machine is a separately excited machine, the magnetic flux is constant. Therefore, we have:

$$\frac{E_{a,1}}{E_{a,2}} = \frac{n_1}{n_2} \Rightarrow \frac{300}{E_{a,2}} = \frac{3000}{2100} \Rightarrow E_{a,2} = 210 \; V \tag{9}$$

The terminal voltage of machine can be calculated by applying voltage division rule as follows:

$$V_{t,2} = \frac{R_L}{R_a + R_L} E_{a,2} = \frac{2.9}{0.1 + 2.9} \times 210 = 203 \; V \tag{10}$$

The terminal current of machine can be calculated as follows:

$$I_{t,2} = \frac{E_{a,2}}{R_a + R_L} = \frac{210}{0.1 + 2.9} = 70 \; A \tag{11}$$

The power delivered to the load can be calculated as follows:

$$P_2 = V_{t,2} I_{t,2} = 203 \times 70 \Rightarrow P_2 = 14210 \; W$$

Choice (3) is the answer.

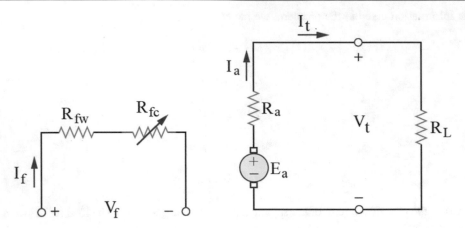

Fig. 10.2 The electrical circuit of solution of problem 10.8

10.9. Based on the information given in the problem, the no-load characteristics of a DC generator at the speed of n_0 are as follows:

E_a in volt	20	50	100	200	300	400
I_f in ampere	0	1	1.6	4	6	9

Moreover, we have:

$$R_a \approx 0 \, \Omega \tag{1}$$

$$V_f = 200 \, V \tag{2}$$

$$V_{t,NL,2} = 200 \, V \tag{3}$$

As we know, the induced armature voltage can be determined by using $E_a = k_a \varphi \omega$. Therefore, we can write:

$$\frac{E_{a,1}}{E_{a,2}} = \frac{\varphi_1 \omega_1}{\varphi_2 \omega_2} \tag{4}$$

However, since the type of machine is separately excited, the magnetic flux is constant. Therefore:

$$\frac{E_{a,1}}{E_{a,2}} = \frac{\omega_1}{\omega_2} = \frac{n_1}{n_2} \tag{5}$$

Therefore, the no-load characteristics of the generator at the speed of $2n_0$ are updated as follows:

$$E_{a,2} = \frac{2n_0}{n_0} E_{a,1} = 2E_{a,1} \tag{3}$$

E_a in volt	40	100	200	400	600	800
I_f in ampere	0	1	1.6	4	6	9

In this problem, since $R_a \approx 0 \, \Omega$, we will have $E_{a, 2} \approx V_{t, NL, 2} = 200 \, V$.

From the no-load characteristics of the generator, for $E_{a, 2} = 200 \, V$, $I_f = 1.6 \, A$ is obtained.

By using Ohm's law and circuit of Fig. 10.3, we have:

$$R_f = R_{fw} + R_{fc} = \frac{V_f}{I_f} = \frac{200}{1.6} = 125 \, \Omega$$

Choice (4) is the answer.

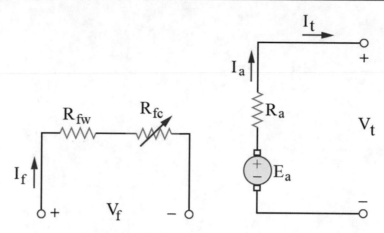

Fig. 10.3 The electrical circuit of solution of problem 10.9

Abstract

In this chapter, the problems concerned with the shunt DC electric generators are solved. In this chapter, the problems are categorized in different levels based on their difficulty levels (easy, normal, and hard) and calculation amounts (small, normal, and large). Additionally, the problems are ordered from the easiest problem with the smallest computations to the most difficult problems with the largest calculations.

11.1. In a 250 V shunt DC generator, the armature winding type is simple lap and the winding includes 360 conductors totally. Herein, the motor is rotating at the speed of 500 rpm, the armature winding current is 25 A, the armature winding resistance is 0.2 Ω, and the voltage drop of each brush is 1 V. Calculate the magnetic flux under each pole.

Difficulty level   ○ Easy   ● Normal   ○ Hard
Calculation amount   ○ Small   ● Normal   ○ Large
1) 83.3 mWb
2) 84.3 mWb
3) 81 mWb
4) 42 mWb

11.2. In a shunt DC generator, the field winding resistance and the armature winding resistance are about 100 Ω and 0 Ω, respectively, and the rated voltage is delivered at the speed of 1000 rpm. Calculate the critical speed if the critical resistance of the field winding is 200 Ω.

Difficulty level   ○ Easy   ○ Normal   ● Hard
Calculation amount   ● Small   ○ Normal   ○ Large
1) 1000 rpm
2) 750 rpm
3) 500 rpm
4) 900 rpm

11.3. In a shunt DC generator, the no-load characteristics equation at the speed of 1200 rpm is as follows.

$$E_a = \frac{200}{1 + I_f}$$

The field winding resistance is 100 Ω. Calculate the output voltage at the speed of 1500 rpm.

Difficulty level   ○ Easy   ○ Normal   ● Hard
Calculation amount   ○ Small   ● Normal   ○ Large
1) 200 V
2) 250 V
3) 232 V
4) 116 V

© The Author(s), under exclusive license to Springer Nature Switzerland AG 2022
M. Rahmani-Andebili, *DC Electric Machines, Electromechanical Energy Conversion Principles, and Magnetic Circuit Analysis*,
https://doi.org/10.1007/978-3-031-08863-6_11

11.4. The field winding resistance and the armature winding resistance of a DC generator are 100 Ω and 1 Ω, respectively. The generator is operated at the rated speed of 2000 rpm and its no-load characteristics are graphically shown in Fig. 11.1. The green color dashed line illustrates the curve of field winding resistance. At what speed the generator is self-excited?

Difficulty level   ○ Easy   ○ Normal   ● Hard
Calculation amount   ○ Small   ● Normal   ○ Large

1) 2000 rpm
2) 1200 rpm
3) 1010 rpm
4) 1000 rpm

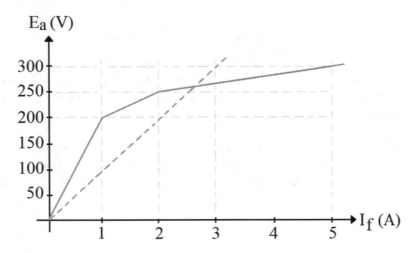

Fig. 11.1 The graph of problem 11.4

11.5. The no-load characteristics of a shunt DC generator are graphically shown in Fig. 11.2. While the armature current is 100 A and the output voltage is 240 V, the load is removed from the generator. In this condition, calculate the output voltage. Herein, assume that the speed of generator remains constant. Moreover, the armature winding resistance is 0.1 Ω.

Difficulty level   ○ Easy   ○ Normal   ● Hard
Calculation amount   ○ Small   ● Normal   ○ Large

1) 59 V
2) 243.33 V
3) 247.67 V
4) 253.33 V

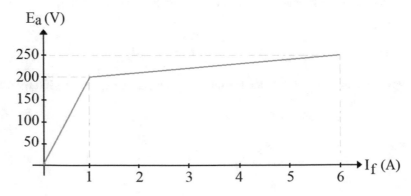

Fig. 11.2 The graph of problem 11.5

11.6. The no-load characteristics of a shunt DC generator are presented by the equation of $E_{a,NL} = 200\sqrt{I_{f,NL}}$. What is the maximum armature current in the no-load condition if the field winding resistance and the armature winding resistance are 200 Ω and 1 Ω, respectively?

Difficulty level   ○ Easy   ○ Normal   ● Hard
Calculation amount   ○ Small   ● Normal   ○ Large

1) 100 A
2) 25 A
3) 50 A
4) 5 A

Solutions of Problems: Shunt DC Electric Generator

12

Abstract

In this chapter, the problems of the 11th chapter are fully solved, in detail, step-by-step, and with different methods.

12.1. Based on the information given in the problem, the winding type of armature is simple lap. In addition, we have:

$$V_t = 250 \ V \tag{1}$$

$$Z = 360 \ V \tag{2}$$

$$n = 500 \ rpm \tag{3}$$

$$I_a = 25 \ A \tag{4}$$

$$R_a = 0.2 \ \Omega \tag{5}$$

$$V_b = 1 \ V \tag{6}$$

Figure 12.1 shows the electrical circuit of the shunt DC generator. By applying KVL in the right-hand side mesh, we have:

$$E_a = V_t + R_a I_a - 2V_b = 250 + (0.2 \times 25) - (2 \times 1) = 253 \ V \tag{7}$$

As we know, the armature voltage of a DC machine can be calculated as follows:

$$E_a = k_a \varphi \omega = \frac{pZ}{2\pi a} \varphi \omega = \frac{pZ}{60a} \varphi n \tag{8}$$

Since the armature includes a simple lap winding, $a = p$:

$$E_a = \frac{Z}{60} \varphi n \tag{9}$$

$$\Rightarrow 253 = \frac{360}{60} \varphi \times 500 \Rightarrow \varphi = 84.3 \ mWb$$

Choice (2) is the answer.

© The Author(s), under exclusive license to Springer Nature Switzerland AG 2022
M. Rahmani-Andebili, *DC Electric Machines, Electromechanical Energy Conversion Principles, and Magnetic Circuit Analysis*,
https://doi.org/10.1007/978-3-031-08863-6_12

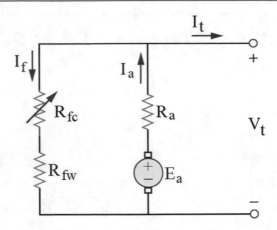

Fig. 12.1 The electrical circuit of solution of problem 12.1

12.2. Based on the information given in the problem, we have:

$$R_f = 100 \; \Omega \tag{1}$$

$$R_a = 0 \; \Omega \tag{2}$$

$$n_{rated} = 1000 \; rpm \tag{3}$$

$$R_{f,c} = 200 \; \Omega \tag{4}$$

As we know, the critical speed is the minimum speed that a generator is self-excited, and it can be determined as follows:

$$n_c = \frac{R_f + R_a}{R_{f,c}} n_{rated} \tag{5}$$

Therefore:

$$n_c = \frac{100 + 0}{200} \times 1000 = 500 \; rpm$$

Choice (3) is the answer.

12.3. Based on the information given in the problem, the no-load characteristics equation of the machine at the speed of 1200 rpm is as follows:

$$E_a = \frac{200}{1 + I_f} \tag{1}$$

Moreover, we have:

$$R_f = 100 \; \Omega \tag{2}$$

$$n_{m1} = 1200 \; rpm \tag{3}$$

$$n_{m2} = 1500 \; rpm \tag{4}$$

The no-load characteristics equation of the machine at the speed of 1500 rpm can be determined as follows:

$$E_a = \frac{200}{1 + I_f} \times \frac{n_{m2}}{n_{m1}} = \frac{200}{1 + I_f} \times \frac{1500}{1200} = \frac{250}{1 + I_f} \tag{5}$$

Now, we need to find the intersection point of the no-load characteristics curve with the line of $E_a = R_f I_f = 100 I_f$ as follows:

$$\frac{250}{1 + I_f} = 100 I_f \Rightarrow I_f^2 + I_f - 2.5 = 0 \Rightarrow I_f = 1.16\, A \tag{6}$$

Therefore:

$$E_a |_{1500\ rpm} = R_f I_f = 100 \times 1.16 = 116\ V$$

Choice (4) is the answer.

12.4. Based on the information given in the problem, the no-load characteristics of the generator are graphically shown in Fig. 12.2. The green color dashed line shows the field winding resistance.

Moreover, we have:

$$R_f = 100\ \Omega \tag{1}$$

$$R_a = 1\ \Omega \tag{2}$$

$$n_{rated} = 2000\ rpm \tag{3}$$

As we know, the minimum speed (critical speed) that a generator is self-excited can be determined by the relation below:

$$n_c = \frac{R_f + R_a}{R_{f,c}} n_{rated} \tag{4}$$

The critical excitation resistance (critical field winding resistance), that is, the ramp of the linear segment of the no-load characteristics curve, can be calculated as follows:

$$R_{f,c} = \frac{200}{1} = 200\ \Omega \tag{5}$$

Solving (1)–(5):

$$n_c = \frac{100 + 1}{200} \times 2000 = 1010\ rpm$$

Choice (3) is the answer.

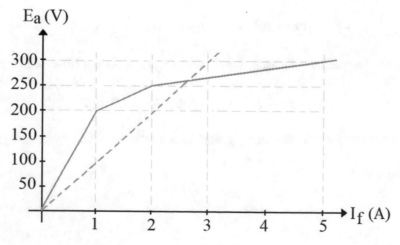

Fig. 12.2 The graph of solution of problem 12.4

12.5. The no-load characteristics of the shunt DC generator are shown in Fig. 12.3.a. In addition, based on the information given in the problem, while the armature current is 100 A and the output voltage is 240 V, the load is removed from the generator. In this condition, the output voltage ($V_{t,\ NL}$) is requested assuming that the speed of generator remains unchanged. Therefore, we have:

$$I_{a,FL} = 100\ A \tag{1}$$

$$V_{t,FL} = 240\ V \tag{2}$$

$$n_{FL} = n_{NL} \tag{3}$$

$$R_a = 0.1\ \Omega \tag{4}$$

Figure 12.3.b shows the electrical circuit of the shunt DC generator. By applying KVL in the right-hand side mesh, we have:

$$E_{a,FL} = V_{t,FL} + R_a I_{a,FL} = 240 + 0.1 \times 100 = 250\ V \tag{5}$$

Since the speed of machine does not change from the full-load condition to the no-load condition, the field winding current in the full-load condition can be calculated by using the no-load characteristics of the machine. Hence:

$$E_{a,FL} = 250\ V \Rightarrow I_{f,FL} = 6\ A \tag{6}$$

By using Ohm's law for the field winding circuit, we have:

$$R_f = R_{fc} + R_{fw} = \frac{V_{t,FL}}{I_{f,FL}} = \frac{240}{6} = 40\ \Omega \tag{7}$$

Therefore, the equation of excitation line (no-load) is as follows:

$$E_{a,NL} = 40 I_{f,NL} \tag{8}$$

On the other hand, the equation of the second line segment of the no-load characteristics of the machine can be calculated as follows:

$$E_{a,NL} = 190 + 10 I_{f,NL} \tag{9}$$

Now, the operating point of the machine in the no-load condition can be determined by finding the intersection point of the curves presented in (8) and (9), as follows:

$$190 + 10 I_{f,NL} = 40 I_{f,NL} \Rightarrow I_{f,NL} = \frac{19}{3}\ A \tag{10}$$

The value of $E_{a,\ NL}$ can be calculated by using any of the equations of (8) and (9) as follows:

$$I_{f,NL} = \frac{19}{3}\ A \Rightarrow E_{a,NL} = 40 \times \frac{19}{3} = 253.33\ V \tag{11}$$

$$\Rightarrow V_{t,NL} \approx E_{a,NL} = 253.33$$

Choice (4) is the answer.

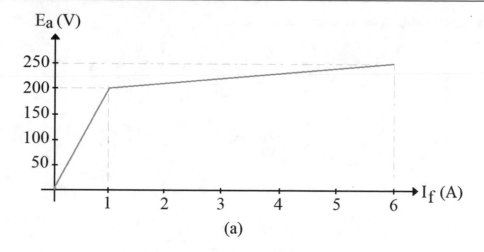

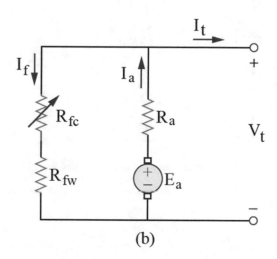

Fig. 12.3 The graph and electrical circuit of solution of problem 12.5

12.6. Based on the information given in the problem, we have:

$$E_{a,NL} = 200\sqrt{I_{f,NL}} \tag{1}$$

$$R_f = 200 \ \Omega \tag{2}$$

$$R_a = 1 \ \Omega \tag{3}$$

Figure 12.4 shows the electrical circuit of the shunt DC generator. By applying KVL in the left-hand side mesh, we have:

$$-E_{a,NL} + R_a I_{a,NL} + R_f I_{f,NL} = 0 \Rightarrow I_{a,NL} = \frac{E_{a,NL} - R_f I_{f,NL}}{R_a} \tag{4}$$

Solving (1)–(4):

$$I_{a,NL} = \frac{200\sqrt{I_{f,NL}} - 200 I_{f,NL}}{1} = 200\left(\sqrt{I_{f,NL}} - I_{f,NL}\right) \tag{5}$$

The maximum value of $I_{a,\,NL}$ can be determined as follows:

$$\frac{dI_{a,NL}}{dI_{f,NL}} = 0 \Rightarrow \frac{1}{2\sqrt{I_{f,NL}}} - 1 = 0 \Rightarrow \sqrt{I_{f,NL}} = \frac{1}{2} \Rightarrow I_{f,NL} = \frac{1}{4}\ A \tag{6}$$

Solving (5) and (6):

$$I_{a,NL} = 200\left(\frac{1}{2} - \frac{1}{4}\right) \Rightarrow I_{a,NL} = 50\ A$$

Choice (3) is the answer.

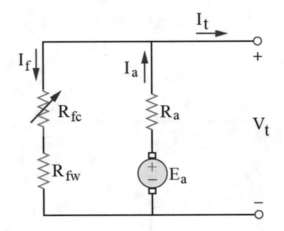

Fig. 12.4 The electrical circuit of solution of problem 12.6

Problems: Series and Compound DC Electric Generators

13

Abstract

In this chapter, the problems concerned with the series and compound DC electric generators are solved. In this chapter, the problems are categorized in different levels based on their difficulty levels (easy, normal, and hard) and calculation amounts (small, normal, and large). Additionally, the problems are ordered from the easiest problem with the smallest computations to the most difficult problems with the largest calculations.

13.1 A 460 V, 100 kW, shunt DC generator has the field current of 2 A in the no-load condition and the field current of 2.65 A in the full-load condition to keep the voltage fixed. How many turns of series winding must be added to each pole so that the voltage remains constant by changing the shunt generator to a short-shunt flat compound generator? The number of turns of the shunt winding is 2000.

Difficulty level ○ Easy ○ Normal ● Hard
Calculation amount ○ Small ● Normal ○ Large
1) 4
2) 6
3) 10
4) 15

13.2 In a series DC generator that its armature winding resistance is about 0.2 Ω, the power of 5 kW at the voltage of 110 V and speed of 1000 rpm is delivered to the load. If the speed is increased to about 1500 rpm, what will be the voltage level for the load of 10 kW? Herein, linear magnetization characteristics are considered for the machine.

Difficulty level ○ Easy ○ Normal ● Hard
Calculation amount ○ Small ● Normal ○ Large
1) 110 V
2) 193 V
3) 220 V
4) 120 V

13.3 The electromotive force (emf) of a long-shunt compound DC generator at the speed of ω can be estimated by relation of $E_a = 0.05\omega + 1.25\omega I_f + 0.01\omega I_a$. The resistance of shunt field winding is 50 Ω which is in series with the field control resistance (R_{fc}). The resistances of armature winding and series field winding are 1 Ω and 1 Ω. Calculate the value of R_{fc} if at the speed of 100 rad/s, the output current and voltage are 10 A and 120 V, respectively.

Difficulty level ○ Easy ○ Normal ● Hard
Calculation amount ○ Small ● Normal ○ Large
1) 34 Ω
2) 59 Ω
3) 69 Ω
4) 85 Ω

© The Author(s), under exclusive license to Springer Nature Switzerland AG 2022
M. Rahmani-Andebili, *DC Electric Machines, Electromechanical Energy Conversion Principles, and Magnetic Circuit Analysis*,
https://doi.org/10.1007/978-3-031-08863-6_13

13.4 The no-load characteristics of a long-shunt compound DC generator at a specific speed are graphically shown in Fig. 13.1. The no-load voltage of the generator is 250 V, and its rated voltage and current are 240 V and 60 A, respectively. The armature winding resistance and the series field winding resistance are 0.25 Ω and 0 Ω, respectively, and the number of shunt field winding is 1000 per pole. Ignore the armature reaction and calculate the number of turns needed for the series field winding for a differential V-I characteristics.

Difficulty level   ○ Easy   ○ Normal   ● Hard
Calculation amount   ○ Small   ● Normal   ○ Large

1) 8
2) 12
3) 14
4) 16

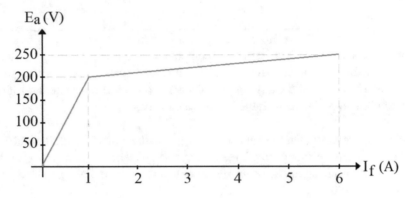

Fig. 13.1 The graph of problem 13.4

Solutions of Problems: Series and Compound DC Electric Generators

14

Abstract

In this chapter, the problems of the 13th chapter are fully solved, in detail, step-by-step, and with different methods.

14.1. Based on the information given in the problem, we have:

$$V_t = 460 \, V \tag{1}$$

$$P = 100 \, kW \tag{2}$$

$$I_{f,NL} = 2 \, A \tag{3}$$

$$I_{f,FL} = 2.65 \, A \tag{4}$$

$$N_{sh} = 2000 \tag{5}$$

The rated terminal current of the shunt DC generator can be calculated as follows (see Fig. 14.1.a):

$$I_t = \frac{P}{V_t} = \frac{100 \times 10^3}{460} = 217.4 \, A \tag{6}$$

Since the shunt DC generator is changed to a short-shunt compound generator, we have (see Fig. 14.1.b):

$$I_s = I_t = 217.4 \, A \tag{7}$$

To change a shunt DC generator to a flat compound DC generator, the number of turns of the series field winding must be as follows:

$$N_s = \frac{\Delta I_f}{I_s} N_{sh} \tag{8}$$

where N_s, N_f, I_s, and ΔI_f are the number of turns of the series field winding, the number of turns of the shunt field winding, the current of series field winding, and the amount change in the field current from the full-load to the no-load condition, respectively.

Solving (8) by considering the value of parameters:

$$N_s = \frac{2.65 - 2}{217.4} \times 2000 \Rightarrow N_s \simeq 6 \, turns$$

Choice (2) is the answer.

© The Author(s), under exclusive license to Springer Nature Switzerland AG 2022
M. Rahmani-Andebili, *DC Electric Machines, Electromechanical Energy Conversion Principles, and Magnetic Circuit Analysis*,
https://doi.org/10.1007/978-3-031-08863-6_14

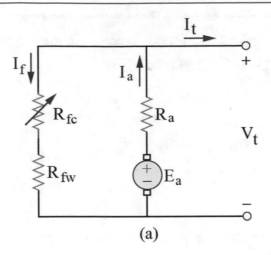

(a)

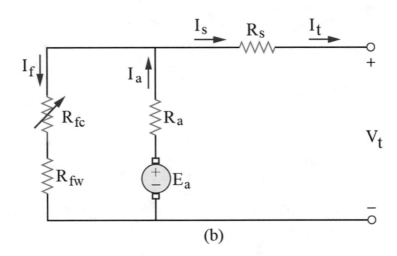

(b)

Fig. 14.1 The electrical circuits of solution of problem 14.1

14.2. Based on the information given in the problem, we have:

$$R_a = 0.2 \ \Omega \tag{1}$$

$$P_1 = 5 \ kW \tag{2}$$

$$V_{t,1} = 110 \ V \tag{3}$$

$$n_1 = 1000 \ rpm \tag{4}$$

$$n_2 = 1500 \ rpm \tag{5}$$

$$P_2 = 10 \ kW \tag{6}$$

Figure 14.2 illustrates the electrical circuit of a series DC generator. The primary and the secondary load currents can be calculated as follows:

$$I_{s,1} = I_{a,1} = I_{t,1} = \frac{P_1}{V_{t,1}} = \frac{5000}{110} = 45.45 \ A \tag{7}$$

$$I_{s,2} = I_{a,2} = I_{t,2} = \frac{P_2}{V_{t,2}} = \frac{10000}{V_{t,2}} \tag{8}$$

By applying KVL in the loop, we have:

$$E_{a,1} = V_{t,1} + R_a I_{a,1} = 110 + (0.2 \times 45.45) = 119.1 \ V \tag{9}$$

$$E_{a,2} = V_{t,2} + R_a I_{a,2} = V_{t,2} + \left(0.2 \times \frac{10000}{V_{t,2}}\right) \tag{10}$$

As we know, the induced armature voltage can be determined by using $E_a = k_a \varphi \omega$. Therefore, we can write:

$$\frac{E_{a,1}}{E_{a,2}} = \frac{\varphi_1 \omega_1}{\varphi_2 \omega_2} = \frac{\varphi_1 n_1}{\varphi_2 n_2} \tag{11}$$

Since linear magnetization characteristics have been considered for the machine, we have:

$$\varphi \propto I_s \tag{12}$$

Therefore:

$$\frac{E_{a,1}}{E_{a,2}} = \frac{I_{s,1} n_1}{I_{s,2} n_2} \tag{13}$$

Solving (13) by considering the value of parameters:

$$\frac{119.1}{V_{t,2} + \frac{2000}{V_{t,2}}} = \frac{1000 \times 45.45}{1500 \times \frac{10000}{V_{t,2}}} \Rightarrow V_{t,2} \approx 193 \ V$$

Choice (2) is the answer.

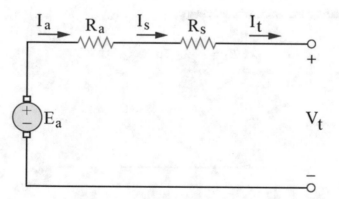

Fig. 14.2 The electrical circuit of solution of problem 14.2

14.3. Based on the information given in the problem, the electromotive force (emf) of a long-shunt compound DC generator at the speed of ω is as follows:

$$E_a = 0.05\omega + 1.25\omega I_f + 0.01\omega I_a \tag{1}$$

$$R_{fw} = 50 \ \Omega \tag{2}$$

$$R_a = 1 \ \Omega \tag{3}$$

$$R_s = 1 \ \Omega \tag{4}$$

$$\omega = 100 \ rad/sec \tag{5}$$

$$I_t = 10\ A \tag{6}$$

$$V_t = 120\ V \tag{7}$$

Figure 14.3 illustrates the electrical circuit of a long-shunt compound DC generator. By applying KVL in the right-hand side mesh and KCL at the top node, we have:

$$E_a = V_t + (R_a + R_s)I_a \tag{8}$$

$$I_a = I_L + I_f = 10 + I_f \tag{9}$$

Solving (8) and (9) and considering the value of parameters:

$$E_a = V_t + (R_a + R_s)(10 + I_f) = 120 + (1+1)(10 + I_f) = 140 + 2I_f \tag{10}$$

Solving (1) and (10) and considering the value of parameters:

$$140 + 2I_f = (0.05 \times 100) + (1.25 \times 100 \times I_f) + 0.01 \times 100 \times (10 + I_f)$$

$$\Rightarrow 140 + 2I_f = 15 + 126I_f \Rightarrow I_f = \frac{125}{124}\ A \tag{11}$$

Applying Ohm's law for the shunt field winding branch:

$$R_{fw} + R_{fc} = \frac{V_t}{I_f} \tag{12}$$

Solving (12) and considering the value of parameters:

$$\Rightarrow R_{fc} = \frac{120}{\frac{125}{124}} - 50 \Rightarrow R_{fc} \approx 69\ \Omega$$

Choice (3) is the answer.

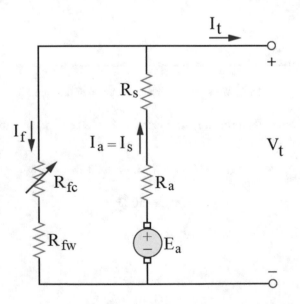

Fig. 14.3 The electrical circuit of solution of problem 14.3

14.4. The no-load characteristics of a long-shunt compound DC generator at a specific speed are shown in Fig. 14.4.a. In addition, the armature reaction is negligible, and a differential V-I characteristics are assumed for the machine.

Moreover, based on the information given in the problem, we have:

$$V_{t,NL} = 250 \, V \tag{1}$$

$$V_{t,FL} = 240 \, V \tag{2}$$

$$I_{t,FL} = 60 \, A \tag{3}$$

$$R_a = 0.25 \, \Omega \tag{4}$$

$$R_s = 0 \, \Omega \tag{5}$$

$$N_{sh} = 1000 \, \text{turns/pole} \tag{6}$$

From the no-load characteristics of machine, the no-load field current can be calculated as follows:

$$E_{a,NL} \approx V_{t,NL} = 250 \, V \Rightarrow I_{f,NL} = 6 \, A \tag{7}$$

Applying Ohm's law for the branch of shunt field winding in the no-load condition:

$$R_f = R_{fw} + R_{fc} = \frac{V_{t,NL}}{I_{f,NL}} = \frac{250}{6} \, \Omega \tag{8}$$

Applying Ohm's law for the branch of shunt field winding in the full-load condition:

$$I_{f,FL} = \frac{V_{t,FL}}{R_f} = \frac{240}{\frac{250}{6}} = 5.76 \, A \tag{9}$$

By applying KVL in the right-hand side mesh of the circuit of Fig. 14.4.b, we have:

$$E_{a,FL} = V_{t,FL} + (R_a + R_s)I_{a,FL} = 240 + (0.25 + 0) \times 60 = 225 \, V \tag{10}$$

From the no-load characteristics of the machine, the full-load field current can be approximately calculated as follows:

$$E_{a,FL} = 255 \, V \xrightarrow{E_a = 10I_f + 190} I_f = 6.5 \, A \tag{11}$$

Now, we need to determine the amount of change in the current of shunt field winding as follows:

$$\Delta I_f = I_{f,FL,new} - I_{f,FL} = 6.5 - 5.76 = 0.74 \, A \tag{12}$$

The number of turns needed for the series field winding with a differential V-I characteristic can be calculated as follows:

$$N_s I_a = N_{sh}\Delta I_f \Rightarrow N_s = N_{sh}\frac{\Delta I_f}{I_a} \tag{13}$$

$$\Rightarrow N_s = 1000 \times \frac{0.74}{60} = 12.33 \Rightarrow N_s \simeq 12 \text{ turns}$$

Choice (2) is the answer.

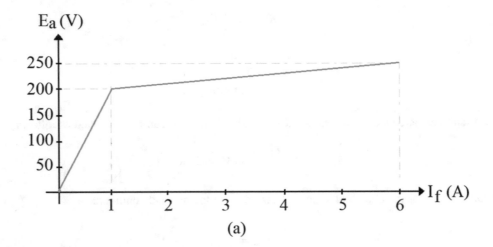

(a)

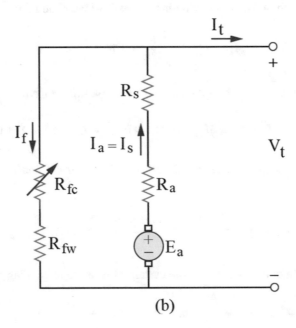

(b)

Fig. 14.4 The graph and electrical circuit of solution of problem 14.4

Problems: Separately Excited and Shunt DC Electric Motors

15

Abstract

In this chapter, the problems concerned with the separately excited and shunt DC electric motors are solved. In this chapter, the problems are categorized in different levels based on their difficulty levels (easy, normal, and hard) and calculation amounts (small, normal, and large). Additionally, the problems are ordered from the easiest problem with the smallest computations to the most difficult problems with the largest calculations.

15.1. In a shunt DC motor, the armature winding resistance is 0.1 Ω. If the input voltage of the motor is 110 V, its input current will be 20 A and the motor will rotate at the speed of 1200 rpm. Now, if the input voltage of the motor is the same as the previous one, its input current is 50 A, and the magnetic flux increases to 10%, what is the new speed of the motor? In this problem, assume that the field circuit draws an ignorable amount of current.

Difficulty level ○ Easy ● Normal ○ Hard
Calculation amount ○ Small ● Normal ○ Large
1) 1121 *rpm*
2) 1283 *rpm*
3) 1061 *rpm*
4) 1358 *rpm*

15.2. Figure 15.1 shows a separately excited DC motor which is operated with a constant load at the speed of n_0. If the position of switch is changed from the position of "1" to "2," what will be the final speed of motor? Herein, assume that magnetic circuit of motor is linear and the ohmic and rotational power losses are ignorable.

Difficulty level ○ Easy ● Normal ○ Hard
Calculation amount ○ Small ● Normal ○ Large
1) n_0
2) $2n_0$
3) $0.5n_0$
4) The information is not enough to determine the speed of motor.

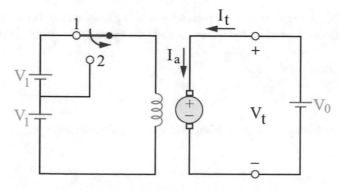

Fig. 15.1 The electrical circuit of problem 15.2

© The Author(s), under exclusive license to Springer Nature Switzerland AG 2022
M. Rahmani-Andebili, *DC Electric Machines, Electromechanical Energy Conversion Principles, and Magnetic Circuit Analysis*,
https://doi.org/10.1007/978-3-031-08863-6_15

15.3. A 240 V shunt DC motor has the no-load speed of 1200 rpm. This motor has the full-load speed of 1320 rpm and draws the armature current of 50 A. The armature winding resistance of the motor is 0.4 Ω. Determine the ratio of magnetic flux in the full-load condition to the no-load one.

Difficulty level   ○ Easy   ● Normal   ○ Hard
Calculation amount   ○ Small   ● Normal   ○ Large
1) $\frac{5}{6}$
2) $\frac{6}{7}$
3) $\frac{7}{6}$
4) $\frac{6}{5}$

15.4. A 250 V, 500 rpm, separately excited DC motor has the armature circuit resistance of 0.3 Ω. The motor draws the armature current of 60 A for the rated torque and rated magnetic flux. If the machine is operated as a generator, what will be its speed with same rated torque and magnetic flux?

Difficulty level   ○ Easy   ● Normal   ○ Hard
Calculation amount   ○ Small   ● Normal   ○ Large
1) 577.6 *rpm*
2) 532.4 *rpm*
3) 564.6 *rpm*
4) 335.7 *rpm*

15.5. In a 220 V shunt DC generator, the armature circuit resistance is 1 Ω. If the full-load current is 20 A, what is the difference between the electromotive forces (emf) in the generating and motoring operations of machine? Herein, ignore the field current.

Difficulty level   ○ Easy   ● Normal   ○ Hard
Calculation amount   ○ Small   ● Normal   ○ Large
1) 20 V
2) 40 V
3) 0 V
4) 50 V

15.6. A 220 V shunt DC machine has the armature winding resistance of 0.5 Ω and the field winding resistance of 110 Ω. If the machine is operated with its rated values, what is the ratio of speed of machine in generating operation to motoring one while the line current is 38 A in both operations.

Difficulty level   ○ Easy   ● Normal   ○ Hard
Calculation amount   ○ Small   ○ Normal   ● Large
1) 1
2) 0.84
3) 2
4) 1.2

15.7. A 10 kW, 200 V, shunt DC generator that has the armature winding resistance of 0.1 Ω and the field winding resistance of 200 Ω is rotating at the rated speed of 750 rpm. Calculate the speed of this machine in motoring operation with the same rated voltage.

Difficulty level   ○ Easy   ● Normal   ○ Hard
Calculation amount   ○ Small   ○ Normal   ● Large
1) 750 *rpm*
2) 713 *rpm*
3) 788 *rpm*
4) 800 *rpm*

15.8. A 5 kW, 200 V, shunt DC generator that has the armature winding resistance of 0.1 Ω and the field winding resistance of 200 Ω is rotating at the rated speed of 1000 rpm. Calculate the speed of the machine if it is operated in motoring mode with the same rated voltage.

Difficulty level ○ Easy ● Normal ○ Hard
Calculation amount ○ Small ○ Normal ● Large

1) 1000 *rpm*
2) 975 *rpm*
3) 925 *rpm*
4) 950 *rpm*

15.9. In a 10 kW, 200 V, shunt DC motor, the armature winding resistance and the field winding resistance are 0.1 Ω and 200 Ω, respectively. The motor is rotating at the speed of 713 rpm with the rated voltage. If the machine is operated as a generator, determine the speed needed for the same field current.

Difficulty level ○ Easy ● Normal ○ Hard
Calculation amount ○ Small ○ Normal ● Large

1) 713 *rpm*
2) 788 *rpm*
3) 800 *rpm*
4) 750 *rpm*

15.10. A 250 V shunt DC machine has the armature winding resistance of 1 Ω and the field winding resistance of 100 Ω. If the machine is supplied by a 250 V power source and draws the line current of 25 A, calculate the ratio of speed of the machine in generating mode to the motoring one.

Difficulty level ○ Easy ● Normal ○ Hard
Calculation amount ○ Small ○ Normal ● Large

1) 1.12
2) 1.02
3) 1.32
4) 0.97

15.11. The no-load characteristics of a shunt DC machine at the speed of 1000 rpm are as follows:

I_f	0	1	2	3	4	5	6	7
E_a	0	50	90	110	120	125	130	135

The machine is supplied by a 240 V power source and rotating with the speed of 2000 rpm in the no-load condition. If its voltage decreases to about 180 V, what will be its speed in the same condition?

Difficulty level ○ Easy ○ Normal ● Hard
Calculation amount ○ Small ● Normal ○ Large

1) 1375 *rpm*
2) 1500 *rpm*
3) 1636 *rpm*
4) 2000 *rpm*

15.12. The magnetization characteristics of a DC shunt machine at the speed of 1150 rpm are illustrated in Fig. 15.2. The machine is operated as a motor and supplied by a 215 V power source. The no-load speed of motor is 1075 rpm. Determine the field winding resistance.

Difficulty level ○ Easy ○ Normal ● Hard
Calculation amount ○ Small ● Normal ○ Large

1) 53.75 Ω
2) 86 Ω
3) 182.73 Ω
4) 195.45 Ω

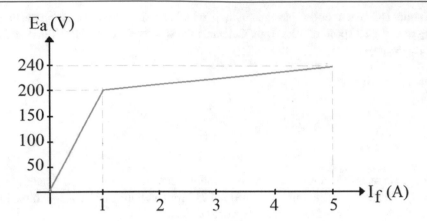

Fig. 15.2 The graph of problem 15.12

15.13. A 200 V, 12 A, 1940 rpm, shunt DC motor has the armature winding resistance of 0.5 Ω. If the applied voltage decreases to about 150 V, calculate the speed of motor if the load torque remains constant, the magnetic circuit is linear, and the field current is negligible.

Difficulty level   ○ Easy   ○ Normal   ● Hard
Calculation amount   ○ Small   ● Normal   ○ Large

1) 1893 *rpm*
2) 1940 *rpm*
3) 1983 *rpm*
4) 1065 *rpm*

15.14. The no-load current of a 220 V shunt DC motor at the speed of 1000 rpm is 2 A. If the full-load current of the machine is 35 A and the armature winding resistance is 0.2 Ω, what is the full-load electromagnetic torque developed? Assume that the magnetic flux remains constant and ignore the current of field circuit.

Difficulty level   ○ Easy   ○ Normal   ● Hard
Calculation amount   ○ Small   ● Normal   ○ Large

1) 71 *N. m*
2) 65 *N. m*
3) 73 *N. m*
4) 77 *N. m*

15.15. The magnetization characteristics of a shunt DC motor are shown in Fig. 15.3. The field winding resistance of the motor is 80 Ω and assume that the developed electromagnetic torque of the motor is proportional to its speed. When the motor is supplied with 240 V, its armature current and speed are 50 A and 1500 rpm, respectively. Now, with a new load, the field winding resistance is increased to about 120 Ω that results in 77 A for the armature current. Calculate the speed of motor in this new condition.

Difficulty level   ○ Easy   ○ Normal   ● Hard
Calculation amount   ○ Small   ● Normal   ○ Large

1) 885.5 *rpm*
2) 960 *rpm*
3) 2100 *rpm*
4) 2310 *rpm*

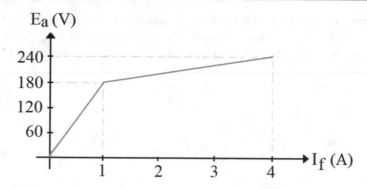

Fig. 15.3. The graph of problem 15.15

15.16. A shunt DC motor that has been connected to a variable mechanical load is supplied by a 200 V power source. When the motor is under load A, its speed and current are 157 rad/s and 20 A. However, when the motor is under load B, its speed and current are 167.5 rad/s and 18 A. Calculate the torque of the loads. Herein, ignore the field current.

Difficulty level   ○ Easy   ○ Normal   ● Hard
Calculation amount   ○ Small   ● Normal   ○ Large

1) $T_A = 13.7\ N.\ m$, $T_B = 15.2\ N.\ m$
2) $T_A = 20.2\ N.\ m$, $T_B = 17.2\ N.\ m$
3) $T_A = 15.2\ N.\ m$, $T_B = 13.7\ N.\ m$
4) $T_A = 10.2\ N.\ m$, $T_B = 8.2\ N.\ m$

15.17. In a shunt DC motor, it is assumed that the magnetic circuit is linear. If the applied voltage is halved in the no-load condition, what will be the ratio of the new speed of the motor to the old one?

Difficulty level   ○ Easy   ○ Normal   ● Hard
Calculation amount   ○ Small   ● Normal   ○ Large

1) 0.5
2) 2
3) 1
4) 0.25

15.18. In a 200 V, 20 A, DC motor, the magnetic flux suddenly increases to about 10%. Calculate the current that this machine will deliver to the electrical grid. Herein, assume that the magnetic circuit of the machine is linear and the armature circuit resistance is 0.1 Ω.

Difficulty level   ○ Easy   ○ Normal   ● Hard
Calculation amount   ○ Small   ● Normal   ○ Large

1) 178 A
2) 79 A
3) 169 A
4) 99 A

15.19. In a shunt DC motor, if the voltage suddenly decreases by about 20%, how much will the armature current increase in a constant load? Herein, ignore the armature winding resistance.

Difficulty level   ○ Easy   ○ Normal   ● Hard
Calculation amount   ○ Small   ● Normal   ○ Large

1) 25%
2) 20%
3) 10%
4) 50%

15.20. When a shunt DC motor is supplied by 200 V, its speed and armature current are 1800 rpm and 100 A. Calculate the no-load speed when its voltage is adjusted at 150 V. The armature winding resistance is negligible, and the magnetization characteristics equation of the machine at the speed of 1800 rpm is as follows:

$$E_a = \frac{500 I_f}{2 + I_f}$$

Difficulty level   ○ Easy   ○ Normal   ● Hard
Calculation amount   ○ Small   ● Normal   ○ Large
1) 1125 rpm
2) 1350 rpm
3) 1620 rpm
4) 1825 rpm

15.21. The armature winding resistance of a 260 V, DC motor, which is loaded with a constant torque, is 0.25 Ω. When the armature current is 40 A, the motor is rotating at 800 rpm. If the magnetic flux under each pole decreases by about 20%, what is the new speed of motor?

Difficulty level   ○ Easy   ○ Normal   ● Hard
Calculation amount   ○ Small   ● Normal   ○ Large
1) 590 rpm
2) 633 rpm
3) 980 rpm
4) 990 rpm

15.22. The magnetization characteristics of a DC shunt motor at a specific speed are illustrated in Fig. 15.4. The motor is under a constant-torque load and its voltage, armature current, armature winding resistance, field winding resistance, and speed are 100 V, 1 A, 0.5 Ω, 100 Ω, and 500 rpm. If the field current is suddenly interrupted, determine the speed of motor.

Difficulty level   ○ Easy   ○ Normal   ● Hard
Calculation amount   ○ Small   ● Normal   ○ Large
1) 475 rpm
2) 1500 rpm
3) 1000 rpm
4) 4750 rpm

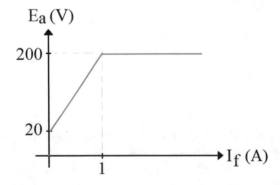

Fig. 15.4 The graph of problem 15.22

15.23. A 40 kW, 500 V, shunt DC motor has the armature winding resistance of 0.1 Ω and the field winding resistance of 250 Ω. The no-load speed of the motor is 1500 rpm that, in this condition, draws 12 A from the power supply. Calculate the developed electromagnetic torque and the speed of motor when it is fully loaded and draws 70 A from the power supply. Herein, assume that the armature reaction decreases by 5% of the magnetic flux under each pole compared to the no-load condition.

Difficulty level   ○ Easy   ○ Normal   ● Hard
Calculation amount   ○ Small   ○ Normal   ● Large

1) 275 $N. m$, 3000 rpm
2) 350 $N. m$, 1820 rpm
3) 275 $N. m$, 1680 rpm
4) 205 $N. m$, 1561 rpm

15.24. In a 220 V shunt DC motor, the armature and field circuit resistances are 0.1 Ω and 100 Ω, respectively, the no-load current is 5 A, and the motor is operated at the speed of 1000 rpm. Calculate the rated output torque if the rated current is 50 A.

Difficulty level   ○ Easy   ○ Normal   ● Hard
Calculation amount   ○ Small   ○ Normal   ● Large

1) 100 $N. m$
2) 94 $N. m$
3) 210 $N. m$
4) 96 $N. m$

15.25. A shunt DC generator has the field winding resistance of 50 Ω and the field control resistance of R_{fc}. The armature winding resistance is negligible. In addition, the electromotive force (mmf) of the machine at the speed of 1200 rpm is as follows:

$$E_a = \begin{cases} 10 + 115 I_f, \, 0 \le I_f \le 2A \\ 240 + 80 \left(I_f - 2 \right), \, I_f \ge 2A \end{cases}$$

The machine is connected to a 240 V supply to be operated as a motor without any load. Calculate the adjusted R_{fc} for the speed of 1000 rpm.

Difficulty level   ○ Easy   ○ Normal   ● Hard
Calculation amount   ○ Small   ○ Normal   ● Large

1) 21 Ω
2) 35 Ω
3) 42 Ω
4) 69 Ω

15.26. The magnetization characteristics of a DC shunt motor at a specific speed are shown in Fig. 15.5. The motor is supplied by a 300 V power source and its speed in the no-load condition is 1200 rpm. The field winding resistance is 100 Ω and the armature winding resistance is negligible. Calculate the speed of motor if 50 Ω resistance and 1 Ω resistance are suddenly added to the filed circuit and armature circuit, respectively, so that the armature current becomes 20 A.

Difficulty level   ○ Easy   ○ Normal   ● Hard
Calculation amount   ○ Small   ○ Normal   ● Large

1) 1173.3 rpm
2) 1120.1 rpm
3) 1169.5 rpm
4) 1069.1 rpm

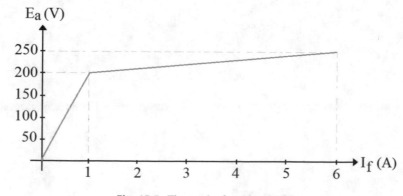

Fig. 15.5 The graph of problem 15.26

Abstract

In this chapter, the problems of the 15th chapter are fully solved, in detail, step-by-step, and with different methods.

16.1. Based on the information given in the problem, we have:

$$R_a = 0.1 \ \Omega \tag{1}$$

$$V_{t,1} = V_{t,2} = 110 \ V \tag{2}$$

$$I_{t,1} = 20 \ A \tag{3}$$

$$n_1 = 1200 \ rpm \tag{4}$$

$$I_{t,2} = 50 \ A \tag{5}$$

$$\varphi_2 = 1.1\varphi_1 \tag{6}$$

$$I_{f,1} = I_{f,2} \approx 0 \ A \tag{7}$$

The electrical circuit of a shunt DC motor is shown in Fig. 16.1. Since $I_f \approx 0 \ A$, we have $I_a = I_t$.

Applying KVL in the right-hand side mesh of the circuit for both conditions:

$$E_{a,1} = V_{t,1} - R_a I_{a,1} = 110 - (0.1 \times 20) = 108 \ V \tag{8}$$

$$E_{a,2} = V_{t,2} - R_a I_{a,2} = 110 - (0.1 \times 50) = 105 \ V \tag{9}$$

As we know, the induced armature voltage of an electric machine can be determined by using $E_a = k_a\varphi\omega$. Therefore, we can write:

$$\frac{E_{a,1}}{E_{a,2}} = \frac{\varphi_1\omega_1}{\varphi_2\omega_2} = \frac{\varphi_1 n_1}{\varphi_2 n_2} \tag{10}$$

$$\Rightarrow \frac{108}{105} = \frac{1200 \times \varphi_1}{n_2 \times (1.1\varphi_1)} \Rightarrow n_2 = 1061 \ rpm$$

Choice (3) is the answer.

© The Author(s), under exclusive license to Springer Nature Switzerland AG 2022
M. Rahmani-Andebili, *DC Electric Machines, Electromechanical Energy Conversion Principles, and Magnetic Circuit Analysis*,
https://doi.org/10.1007/978-3-031-08863-6_16

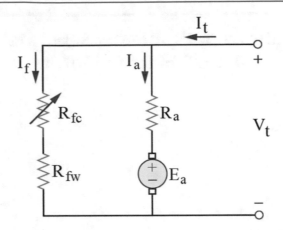

Fig. 16.1   The electrical circuit of solution of problem 16.1

16.2. Based on the information given in the problem, the magnetic circuit of the motor is linear and the ohmic and rotational power losses are ignorable.

Moreover, we have:

$$n_1 = n_0 \tag{1}$$

$$V_{f1} = 2V_1 \tag{2}$$

$$V_{f2} = V_1 \tag{3}$$

$$V_t = V_0 \tag{4}$$

As we know, the induced armature voltage of an electric machine can be determined by using $E_a = k_a \varphi \omega_m$. Therefore, we can write:

$$\frac{E_{a,2}}{E_{a,1}} = \frac{\varphi_2 \omega_2}{\varphi_1 \omega_1} = \frac{\varphi_2 n_2}{\varphi_1 n_1} \tag{5}$$

Since $I_f = \frac{V_f}{R_f}$ and $\varphi \propto I_f$ (linear magnetic circuit), we can conclude that:

$$\varphi \propto V_f \tag{6}$$

Therefore:

$$\frac{E_{a,2}}{E_{a,1}} = \frac{n_2 V_{f,2}}{n_1 V_{f,1}} \tag{7}$$

$$\Rightarrow \frac{V_0}{V_0} = \frac{n_2 \times V_1}{n_0 \times 2V_1} \Rightarrow n_2 = 2n_0$$

Choice (2) is the answer.

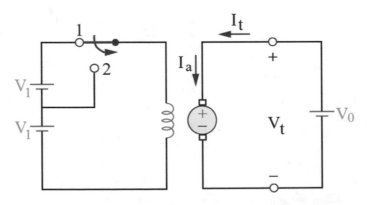

Fig. 16.2   The electrical circuit of solution of problem 16.2

16.3. Based on the information given in the problem, we have:

$$V_t = 240 \tag{1}$$

$$n_{NL} = 1200 \; rpm \tag{2}$$

$$n_{FL} = 1320 \; rpm \tag{3}$$

$$I_a = 50 \, A \tag{4}$$

$$R_a = 0.4 \, \Omega \tag{5}$$

In the no-load condition, we have:

$$E_{a,NL} \simeq V_t = 240 \, V \tag{6}$$

Figure 16.3 shows the electrical circuit of a shunt DC motor. Applying KVL in the right-hand side mesh of the circuit for the full-load condition:

$$E_{a,FL} = V_t - (R_a I_a) = 240 - (0.4 \times 50) = 220 \, V \tag{7}$$

As we know, the emf of an electric machine can be determined by using $E_a = k_a \varphi \omega$. Therefore, we have:

$$\frac{E_{a,FL}}{E_{a,NL}} = \frac{\varphi_{FL} \omega_{FL}}{\varphi_{NL} \omega_{NL}} = \frac{\varphi_{FL} n_{FL}}{\varphi_{NL} n_{NL}} \tag{8}$$

$$\Rightarrow \frac{220}{240} = \frac{\varphi_{FL}}{\varphi_{NL}} \times \frac{1320}{1200} \Rightarrow \frac{\varphi_{FL}}{\varphi_{NL}} = \frac{5}{6}$$

Choice (1) is the answer.

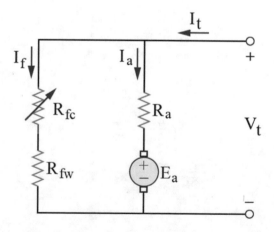

Fig. 16.3 The electrical circuit of solution of problem 16.3

16.4. Based on the information given in the problem, we have:

$$V_{t,g} = V_{t,m} = 250 \ V \tag{1}$$

$$n_m = 500 \ rpm \tag{2}$$

$$R_a = 0.3 \ \Omega \tag{3}$$

$$I_{a,m} = 60 \ A \tag{4}$$

$$T_g = T_m \tag{5}$$

$$\varphi_g = \varphi_m \tag{6}$$

From (4)–(6) and considering $T = k_a \varphi I_a$, we can conclude that:

$$I_{a,g} = 60 \ A \tag{7}$$

Figure 16.4.a and Fig. 16.4.b show the electrical circuit of a separately excited DC machine in motoring and generating modes, respectively.

Applying KVL in the right-hand side mesh of the circuit of Fig. 16.4.a:

$$E_{a,m} = V_{t,m} - R_a I_{a,m} = 250 - (0.3 \times 60) = 232 \ V \tag{8}$$

Applying KVL in the right-hand side mesh of the circuit of Fig. 16.4.b:

$$E_{a,g} = V_{t,g} + R_a I_{a,g} = 250 + (0.3 \times 60) = 268 \ V \tag{9}$$

As we know, the induced armature voltage of an electric machine can be determined by using $E_a = k_a \varphi \omega_m$. Therefore, we can write:

$$\frac{E_{a,m}}{E_{a,g}} = \frac{\varphi_m \omega_m}{\varphi_g \omega_g} = \frac{\varphi_m n_m}{\varphi_g n_g} \tag{10}$$

Solving (6) and (10):

$$\frac{E_{a,m}}{E_{a,g}} = \frac{n_m}{n_g} \tag{11}$$

$$\Rightarrow \frac{232}{268} = \frac{500}{n_g} \Rightarrow n_g = 577.6 \ rpm$$

Choice (1) is the answer.

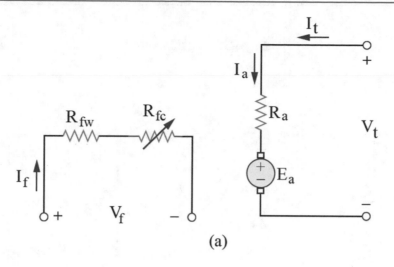

(a)

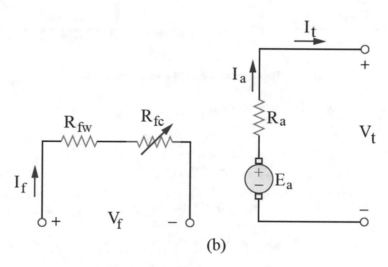

(b)

Fig. 16.4 The electrical circuits of solution of problem 16.4

16.5. Based on the information given in the problem, we have:

$$V_t = 220 \, V \tag{1}$$

$$R_a = 1 \, \Omega \tag{2}$$

$$I_{t,FL} = 20 \, A \tag{3}$$

$$I_f \approx 0 \, A \tag{4}$$

Figure 16.5 shows the electrical circuits of a shunt DC generator and motor. From (4), we have:

$$I_a = I_t \tag{5}$$

Applying KVL in the right-hand side mesh of the circuit of Fig. 16.5.a:

$$E_{a,g} = V_t + R_a I_{a,g} = 220 + (1 \times 20) = 240 \, V \tag{6}$$

Applying KVL in the right-hand side mesh of the circuit of Fig. 16.5.b:

$$E_{a,m} = V_t - R_a I_{a,m} = 220 - (1 \times 20) = 200 \, V \tag{7}$$

Therefore:

$$\Delta E_a = E_{a,g} - E_{a,m} = 40 \ V$$

Choice (2) is the answer.

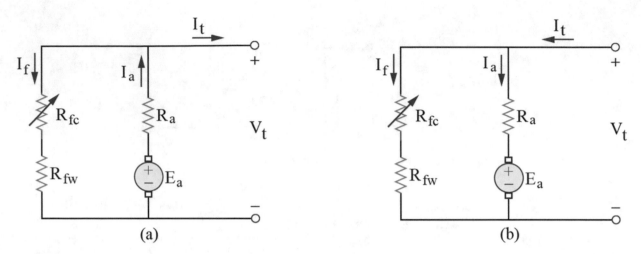

Fig. 16.5 The electrical circuits of solution of problem 16.5

16.6. Based on the information given in the problem, we have:

$$V_{t,g} = V_{t,m} = 220 \ V \tag{1}$$

$$R_a = 0.5 \ \Omega \tag{2}$$

$$R_f = 110 \ \Omega \tag{3}$$

$$I_{t,m} = I_{t,g} = 38 \ A \tag{4}$$

Figure 16.6.a and Fig. 16.6.b show the electrical circuit of a shunt DC machine in generating and motoring operations, respectively.

Applying Ohm's law for the branch of field winding in Fig. 16.6.a:

$$I_{f,g} = I_{f,m} = \frac{V_{t,g}}{R_{fc} + R_{fw}} = \frac{V_{t,g}}{R_f} = \frac{220}{110} = 2 \ A \tag{5}$$

Applying KCL at the top node of the circuit of Fig. 16.6.a:

$$I_{a,g} = I_{t,g} + I_{f,g} = 38 + 2 = 40 \ A \tag{6}$$

Applying KVL in the right-hand side mesh of the circuit of Fig. 16.6.a:

$$E_{a,g} = V_{t,g} + R_a I_{a,g} = 220 + (0.5 \times 40) = 240 \ V \tag{7}$$

Applying KCL at the top node of the circuit of Fig. 16.6.b:

$$I_{a,m} = I_{t,m} - I_{f,m} = 38 - 2 = 36 \ A \tag{8}$$

Applying KVL in the right-hand side mesh of the circuit of Fig. 16.6.b:

$$E_{a,m} = V_{t,m} - R_a I_{a,m} = 220 - (0.5 \times 36) = 202 \ V \tag{9}$$

As we know, the induced armature voltage of an electric machine can be determined by using $E_a = k_a \varphi \omega$. Therefore, we can write:

$$\frac{E_{a,g}}{E_{a,m}} = \frac{\varphi_g \omega_g}{\varphi_m \omega_m} = \frac{\varphi_g n_g}{\varphi_m n_m} \tag{10}$$

Since $I_{f,g} = I_{f,m}$, we have:

$$\varphi_g = \varphi_m \tag{11}$$

Therefore:

$$\frac{E_{a,g}}{E_{a,m}} = \frac{n_g}{n_m} \tag{12}$$

$$\Rightarrow \frac{240}{202} = \frac{n_g}{n_m} \Rightarrow \frac{n_g}{n_m} = 1.2$$

Choice (4) is the answer.

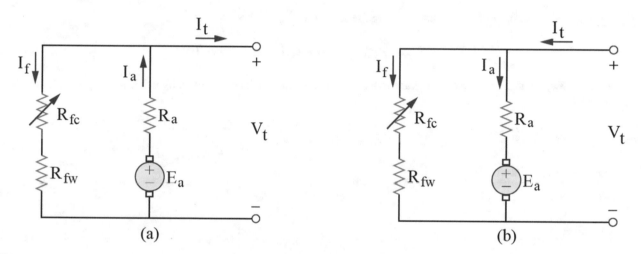

Fig. 16.6 The electrical circuits of solution of problem 16.6

16.7. Based on the information given in the problem, we have:

$$P_g = P_m = 10 \ kW \tag{1}$$

$$V_{t,g} = V_{t,m} = 200 \ V \tag{2}$$

$$R_a = 0.1 \ \Omega \tag{3}$$

$$R_f = 200 \ \Omega \tag{4}$$

$$n_g = 750 \ rpm \tag{5}$$

Figure 16.7.a and Fig. 16.7.b show the electrical circuit of a shunt DC machine in generating and motoring operations, respectively.

Applying Ohm's law for the branch of field winding in Fig. 16.7.a:

$$I_{f,g} = I_{f,m} = \frac{V_{t,g}}{R_{fc} + R_{fw}} = \frac{V_{t,g}}{R_f} = \frac{200}{200} = 1 \ A \tag{6}$$

The rated terminal current of the machine can be calculated as follows.

$$I_{t,g} = I_{t,m} = \frac{P_g}{V_{t,g}} = \frac{10 \times 10^3}{200} = 50 \, A \tag{7}$$

Applying KCL at the top node of the circuit of Fig. 16.7.a:

$$I_{a,g} = I_{t,g} + I_{f,g} = 50 + 1 = 51 \, A \tag{8}$$

Applying KVL in the right-hand side mesh of the circuit of Fig. 16.7.a:

$$E_{a,g} = V_{t,g} + R_a I_{a,g} = 200 + (0.1 \times 51) = 205.1 \, V \tag{9}$$

Applying KCL at the top node of the circuit of Fig. 16.7.b:

$$I_{a,m} = I_{t,m} - I_{f,m} = 50 - 1 = 49 \, A \tag{10}$$

Applying KVL in the right-hand side mesh of the circuit of Fig. 16.7.b:

$$E_{a,m} = V_{t,m} - R_a I_{a,m} = 200 - (0.1 \times 49) = 195.1 \, V \tag{11}$$

As we know, the induced armature voltage of an electric machine can be determined by using $E_a = k_a \varphi \omega$. Therefore, we can write:

$$\frac{E_{a,g}}{E_{a,m}} = \frac{\varphi_g \omega_g}{\varphi_m \omega_m} = \frac{\varphi_g n_g}{\varphi_m n_m} \tag{12}$$

Since $I_{f,\,g} = I_{f,\,m}$, we have:

$$\varphi_g = \varphi_m \tag{13}$$

Therefore:

$$\frac{E_{a,g}}{E_{a,m}} = \frac{n_g}{n_m} \tag{14}$$

$$\Rightarrow \frac{205.1}{195.1} = \frac{750}{n_m} \Rightarrow n_m \approx 713 \, rpm$$

Choice (2) is the answer.

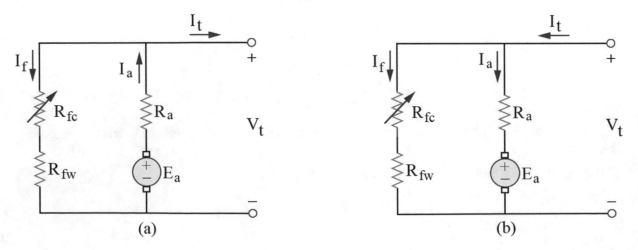

Fig. 16.7 The electrical circuits of solution of problem 16.7

16.8. Based on the information given in the problem, we have:

$$P_g = P_m = 5 \ kW \tag{1}$$

$$V_{t,g} = V_{t,m} = 200 \ V \tag{2}$$

$$R_a = 0.1 \ \Omega \tag{3}$$

$$R_f = 200 \ \Omega \tag{4}$$

$$n_g = 1000 \ rpm \tag{5}$$

Figure 16.8.a and Fig. 16.8.b show the electrical circuit of a shunt DC machine in generating and motoring operations, respectively.

Applying Ohm's law for the branch of field winding in Fig. 16.8.a:

$$I_{f,g} = I_{f,m} = \frac{V_{t,g}}{R_{fc} + R_{fw}} = \frac{V_{t,g}}{R_f} = \frac{200}{200} = 1 \ A \tag{6}$$

The rated terminal current of the machine can be calculated as follows:

$$I_{t,g} = I_{t,m} = \frac{P_g}{V_{t,g}} = \frac{5000}{200} = 25 \ A \tag{7}$$

Applying KCL at the top node of the circuit of Fig. 16.8.a:

$$I_{a,g} = I_{t,g} + I_{f,g} = 25 + 1 = 26 \ A \tag{8}$$

Applying KVL in the right-hand side mesh of the circuit of Fig. 16.8.a:

$$E_{a,g} = V_{t,g} + R_a I_{a,g} = 200 + (0.1 \times 26) = 202.6 \ V \tag{9}$$

Applying KCL at the top node of the circuit of Fig. 16.8.b:

$$I_{a,m} = I_{t,m} - I_{f,m} = 25 - 1 = 24 \ A \tag{10}$$

Applying KVL in the right-hand side mesh of the circuit of Fig. 16.8.b:

$$E_{a,m} = V_{t,m} - R_a I_{a,m} = 200 - (0.1 \times 24) = 197.6 \ V \tag{11}$$

As we know, the induced armature voltage of an electric machine can be determined by using $E_a = k_a \varphi \omega$. Therefore, we can write:

$$\frac{E_{a,g}}{E_{a,m}} = \frac{\varphi_g \omega_g}{\varphi_m \omega_m} = \frac{\varphi_g n_g}{\varphi_m n_m} \tag{12}$$

Since $I_{f,\ g} = I_{f,\ m}$, we have:

$$\varphi_g = \varphi_m \tag{13}$$

Therefore:

$$\frac{E_{a,g}}{E_{a,m}} = \frac{n_g}{n_m} \tag{14}$$

$$\Rightarrow \frac{202.6}{197.6} = \frac{1000}{n_m} \Rightarrow n_m \approx 975 \; rpm$$

Choice (2) is the answer.

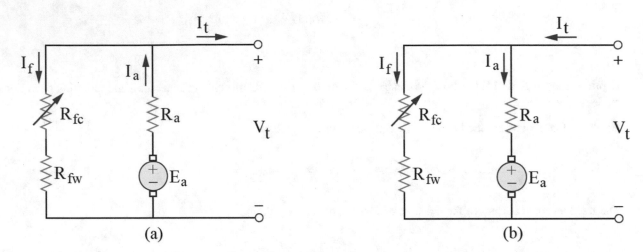

Fig. 16.8 The electrical circuits of solution of problem 16.8

16.9. Based on the information given in the problem, we have:

$$P_g = P_m = 10 \; kW \tag{1}$$

$$V_{t,g} = V_{t,m} = 200 \; V \tag{2}$$

$$R_a = 0.1 \; \Omega \tag{3}$$

$$R_f = 200 \; \Omega \tag{4}$$

$$n_m = 713 \; rpm \tag{5}$$

Figure 16.9.a and Fig. 16.9.b show the electrical circuit of a shunt DC machine in generating and motoring modes, respectively.

The rated terminal current of the machine can be calculated as follows:

$$I_{t,g} = I_{t,m} = \frac{P_g}{V_{t,g}} = \frac{10000}{200} = 50 \; A \tag{6}$$

Applying Ohm's law for the branch of field winding in Fig. 16.9.a:

$$I_{f,g} = I_{f,m} = \frac{V_{t,g}}{R_{fc} + R_{fw}} = \frac{V_{t,g}}{R_f} = \frac{200}{200} = 1 \; A \tag{7}$$

Applying KCL at the top node of the circuit of Fig. 16.9.a:

$$I_{a,g} = I_{t,g} + I_{f,g} = 50 + 1 = 51 \; A \tag{8}$$

Applying KVL in the right-hand side mesh of the circuit of Fig. 16.9.a:

$$E_{a,g} = V_{t,g} + R_a I_{a,g} = 200 + (0.1 \times 51) \simeq 205 \ V \tag{9}$$

Applying KCL at the top node of the circuit of Fig. 16.9.b:

$$I_{a,m} = I_{t,m} - I_{f,m} = 50 - 1 = 49 \ A \tag{10}$$

Applying KVL in the right-hand side mesh of the circuit of Fig. 16.9.b:

$$E_{a,m} = V_{t,m} - R_a I_{a,m} = 200 - (0.1 \times 49) \simeq 195 \ V \tag{11}$$

As we know, the induced armature voltage of an electric machine can be determined by using $E_a = k_a \varphi \omega$. Therefore, we can write:

$$\frac{E_{a,m}}{E_{a,g}} = \frac{\varphi_m \omega_m}{\varphi_g \omega_g} = \frac{\varphi_m n_m}{\varphi_g n_g} \tag{12}$$

Since $I_{f,\ g} = I_{f,\ m}$, we have:

$$\varphi_g = \varphi_m \tag{13}$$

Therefore:

$$\frac{E_{a,m}}{E_{a,g}} = \frac{n_m}{n_g} \tag{14}$$

$$\Rightarrow \frac{195}{205} = \frac{713}{n_m} \Rightarrow n_m = 750 \ rpm$$

Choice (4) is the answer.

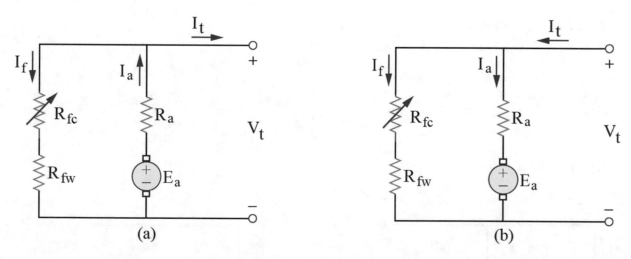

Fig. 16.9 The electrical circuits of solution of problem 16.9

16.10. Based on the information given in the problem, we have:

$$V_{t,g} = V_{t,m} = 250 \ V \tag{1}$$

$$R_a = 1 \ \Omega \tag{2}$$

$$R_f = 100 \ \Omega \tag{3}$$

$$I_{t,m} = I_{t,g} = 25 \ A \tag{4}$$

Figure 16.10.a and Fig. 16.10.b show the electrical circuit of shunt DC machine in generating and motoring operations, respectively.

Applying Ohm's law for the branch of field winding in Fig. 16.10.a:

$$I_{f,g} = I_{f,m} = \frac{V_{t,g}}{R_{fc} + R_{fw}} = \frac{V_{t,g}}{R_f} = \frac{250}{100} = 2.5 \, A \tag{5}$$

Applying KCL at the top node of the circuit of Fig. 16.10.a:

$$I_{a,g} = I_{t,g} + I_{f,g} = 25 + 2.5 = 27.5 \tag{6}$$

Applying KVL in the right-hand side mesh of the circuit of Fig. 16.10.a:

$$E_{a,g} = V_{t,g} + R_a I_{a,g} = 250 + 0.1 \times 27.5 = 252.75 \, V \tag{7}$$

Applying KCL at the top node of the circuit of Fig. 16.10.b:

$$I_{a,m} = I_{t,m} - I_{f,m} = 25 - 2.5 = 22.5 \tag{8}$$

Applying KVL in the right-hand side mesh of the circuit of Fig. 16.10.b:

$$E_{a,m} = V_{t,m} - R_a I_{a,m} = 250 - 0.1 \times 22.5 = 247.75 \, V \tag{9}$$

As we know, the induced armature voltage of an electric machine can be determined by using $E_a = k_a \varphi \omega$. Therefore, we can write:

$$\frac{E_{a,g}}{E_{a,m}} = \frac{\varphi_g \omega_g}{\varphi_m \omega_m} = \frac{\varphi_g n_g}{\varphi_m n_m} \tag{10}$$

Since $I_{f,\, m} = I_{f,\, g}$, we have:

$$\varphi_m = \varphi_g \tag{11}$$

Solving (10) and (11):

$$\Rightarrow \frac{252.75}{247.75} = \frac{n_g}{n_m} \Rightarrow \frac{n_g}{n_m} = 1.02$$

Choice (2) is the answer.

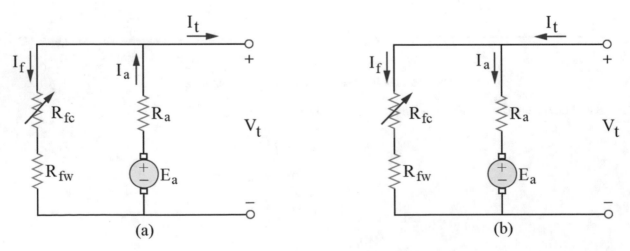

(a)   (b)

Fig. 16.10 The electrical circuits of solution of problem 16.10

16.11. Based on the information given in the problem, we know that:

$$V_{t,1} = 240 \ V \tag{1}$$

$$n_{NL,1} = 2000 \ rpm \tag{2}$$

$$V_{t,2} = 180 \ V \tag{3}$$

Moreover, the no-load characteristics of the machine at the speed of 1000 rpm are as follows:

I_f	0	1	2	3	4	5	6	7
E_a at 1000 rpm	0	50	90	110	120	125	130	135

The no-load characteristics of the machine at the speed of 2000 rpm can be updated as follows:

I_f	0	1	2	3	4	5	6	7
E_a at 2000 rpm	0	100	180	220	240	250	260	270

Now, we can extract the field winding current from the no-load characteristics of the machine at the speed of 2000 rpm as follows:

$$\widehat{E}_{a,1}|_{2000 \ rpm} \approx V_{t,1} = 240 \ V \Rightarrow I_{f,1} = 4 \ A \tag{4}$$

Figure 16.11 shows the electrical circuit of a shunt DC motor. Applying Ohm's law for the branch of field winding (the first condition):

$$R_f = \frac{V_{t,1}}{I_{f,1}} = \frac{240}{4} = 60 \ \Omega \tag{5}$$

Applying Ohm's law for the branch of field winding (the second condition):

$$I_{f,2} = \frac{V_{t,2}}{R_f} = \frac{180}{60} = 3 \ A \tag{6}$$

Now, we can extract the emf of the motor from its no-load characteristics at the speed of 2000 rpm as follows:

$$I_{f,2} = 3 \ A \Rightarrow \widehat{E}_{a,2}|_{2000 \ rpm} = 220 \ V \tag{7}$$

As we know, the induced armature voltage of an electric machine can be determined by using $E_a = k_a \varphi \omega$. Therefore, we can write:

$$\frac{E_{a,1}}{E_{a,2}} = \frac{\varphi_1 \omega_1}{\varphi_2 \omega_2} = \frac{\varphi_1 n_1}{\varphi_2 n_2} \tag{8}$$

$$\Rightarrow \frac{\widehat{E}_{a,1}}{\widehat{E}_{a,2}}|_{2000 \ rpm} = \frac{\varphi_1}{\varphi_2} \Rightarrow \frac{\varphi_1}{\varphi_2} = \frac{240}{220} \tag{9}$$

Now, for both no-load conditions, we have the relation below:

$$\frac{V_{t,1}}{V_{t,2}} \approx \frac{E_{a,1}}{E_{a,2}} = \frac{\varphi_1}{\varphi_2} \frac{n_{NL,1}}{n_{NL,2}} \tag{10}$$

$$\Rightarrow \frac{180}{240} = \frac{240}{220} \times \frac{2000}{n_{NL,2}} \Rightarrow n_{NL,2} \approx 1636 \ rpm$$

Choice (3) is the answer.

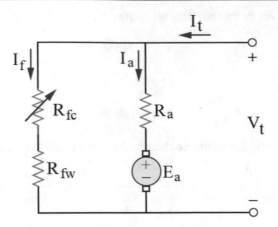

Fig. 16.11 The electrical circuit of solution of problem 16.11

16.12. The magnetization characteristics of the shunt DC motor at the speed of 1150 rpm are illustrated in Fig. 16.12. In addition, based on the information given in the problem, we have:

$$V_t = 215 \ V \tag{1}$$

$$n_{FL} = 1150 \ rpm \tag{2}$$

$$n_{NL} = 1075 \ rpm \tag{3}$$

The equation of the second segment of the magnetization characteristics of machine at the speed of 1150 rpm can be determined as follows:

$$E_{a,FL} = 10I_f + 190 \tag{4}$$

Moreover, the equation of the second segment of the magnetization characteristics of machine at the speed of 1075 rpm can be updated as follows:

$$E_{a,NL} = \left(10I_f + 190\right)\frac{1075}{1150} \tag{5}$$

As we know, in the no-load condition, we have:

$$E_{a,NL} \simeq V_t = 215 \ V \tag{6}$$

Solving (5) and (6):

$$215 = \left(10I_f + 190\right) \times \frac{1075}{1150} \Rightarrow I_f = 4 \ A \tag{7}$$

Applying Ohm's law for the branch of field winding:

$$R_f = R_{fw} + R_{fc} = \frac{V_t}{I_f} = \frac{215}{4} \Rightarrow R_f = 53.75 \ \Omega$$

Choice (1) is the answer.

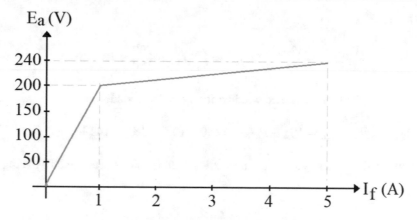

Fig. 16.12 The graph and electrical circuit of solution of problem 16.12

16.13. Based on the information given in the problem, we have:

$$V_{t,1} = 200 \ V \tag{1}$$

$$I_{t,1} = 12 \ A \tag{2}$$

$$n_1 = 1940 \ rpm \tag{3}$$

$$R_a = 0.5 \ \Omega \tag{4}$$

$$V_{t,2} = 150 \ V \tag{5}$$

$$T_{load,2} = T_{load,1} \tag{6}$$

$$\varphi \propto I_f \tag{7}$$

$$I_f \approx 0 \tag{8}$$

Since $I_f \approx 0$, we have:

$$I_a = I_t \tag{9}$$

Figure 16.13 shows the electrical circuit of a shunt DC motor. Applying KVL in the right-hand side mesh of the circuit:

$$E_{a,1} = V_{t,1} - R_a I_{a,1} = 200 - (0.5 \times 12) = 194 \ V \tag{10}$$

From (6), we can conclude:

$$T_{e,2} = T_{e,1} \tag{11}$$

As we know, the electromagnetic torque (T_e) of an electric machine can be determined by using $T_e = k_a \varphi I_a$. Therefore, from (11), we can write:

$$k_a \varphi_2 I_{a,2} = k_a \varphi_1 I_{a,1} \Rightarrow \varphi_2 I_{a,2} = \varphi_1 I_{a,1} \overset{(7)}{\Longrightarrow} I_{f,2} I_{a,2} = I_{f,1} I_{a,1} \tag{12}$$

Applying Ohm's law for the branch of field winding for the first condition:

$$I_f = \frac{V_t}{R_{fc} + R_{fw}} = \frac{V_t}{R_f} \tag{13}$$

Solving (9), (12), and (13):

$$\frac{V_{t,1}}{R_f}I_{a,1} = \frac{V_{t,2}}{R_f}I_{a,2} \Rightarrow I_{a,2} = \frac{V_{t,1}}{V_{t,2}}I_{a,1} = \frac{200}{150} \times 12 = 16\,A \tag{14}$$

Applying Ohm's law for the branch of field winding for the second condition:

$$E_{a,2} = V_{t,2} - R_aI_{a,2} = 150 - (0.5 \times 16) = 142\,V \tag{15}$$

As we know, the induced armature voltage of an electric machine can be determined by using $E_a = k_a\varphi\omega$. Therefore, we can write:

$$\frac{E_{a,1}}{E_{a,2}} = \frac{\varphi_1\omega_1}{\varphi_2\omega_2} = \frac{\varphi_1 n_1}{\varphi_2 n_2} \tag{16}$$

Solving (7) and (13):

$$\varphi \propto V_t \tag{17}$$

Solving (16) and (17):

$$\frac{E_{a,1}}{E_{a,2}} = \frac{V_{t,1}n_1}{V_{t,2}n_2} \tag{18}$$

$$\Rightarrow \frac{194}{142} = \frac{200}{150} \times \frac{1940}{n_2} \Rightarrow n_2 \approx 1893\,rpm$$

Choice (1) is the answer.

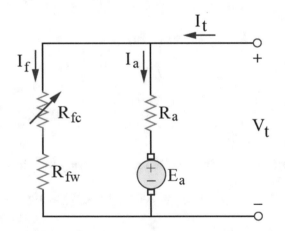

Fig. 16.13 The electrical circuit of solution of problem 16.13

16.14. Based on the information given in the problem, we have:

$$V_t = 220\,V \tag{1}$$

$$n_{NL} = 1000\,rpm \tag{2}$$

$$I_{t,NL} = 2\,A \tag{3}$$

$$I_{t,FL} = 35\,A \tag{4}$$

$$R_a = 0.2 \ \Omega \tag{5}$$

$$\varphi_{FL} = \varphi_{NL} \tag{6}$$

$$I_f \approx 0 \ A \tag{7}$$

Since $I_f \approx 0 \ A$, we have $I_a \approx I_t$.

Figure 16.14 shows the electrical circuit of a shunt DC motor. Applying KVL in the right-hand side mesh of the circuit:

$$E_{a,NL} = V_t - R_a I_{a,NL} = 220 - (0.2 \times 2) = 219.6 \ V \tag{8}$$

The developed electromagnetic torque in the no-load condition can be calculated as follows:

$$T_{e,NL} = \frac{P_{e,NL}}{\omega_{NL}} = \frac{E_{a,NL} I_{a,NL}}{n_{NL} \frac{2\pi}{60}} = \frac{219.6 \times 2}{1000 \times \frac{2\pi}{60}} = 4.2 \ N.m \tag{9}$$

As we know, the electromagnetic torque of an electric machine can be determined by using $T_e = k_a \varphi I_a$. Therefore, we can write:

$$\frac{T_{e,FL}}{T_{e,NL}} = \frac{\varphi_{FL} I_{a,FL}}{\varphi_{NL} I_{a,NL}} \tag{10}$$

Since the terminal voltage is constant, we have:

$$I_{f,FL} = I_{f,NL} \Rightarrow \varphi_{FL} = \varphi_{NL} \tag{11}$$

Therefore:

$$\frac{T_{e,FL}}{T_{e,NL}} = \frac{I_{a,FL}}{I_{a,NL}} \tag{12}$$

$$\Rightarrow \frac{T_{e,FL}}{4.2} = \frac{35}{2} \Rightarrow T_{e,FL} \approx 73 \ N.m$$

Choice (3) is the answer.

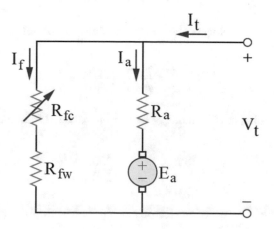

Fig. 16.14 The electrical circuit of solution of problem 16.14

16.15. The magnetization characteristics of the shunt DC motor are illustrated in Fig. 16.15.a. In addition, based on the information given in the problem, we have:

$$T_e \propto n \tag{1}$$

$$V_t = 240 \ V \tag{2}$$

$$R_{f,1} = 80 \ \Omega \tag{3}$$

$$I_{a,1} = 50 \ A \tag{4}$$

$$n_1 = 1500 \ rpm \tag{5}$$

$$R_{f,2} = 120 \ \Omega \tag{6}$$

$$I_{a,2} = 77 \ A \tag{7}$$

Figure 16.15.b shows the electrical circuit of a shunt DC motor. Applying Ohm's law for the branch of field winding:

$$I_{f,1} = \frac{V_t}{R_{fc,1} + R_{fw}} = \frac{V_t}{R_{f,1}} = \frac{240}{80} = 3 \ A \tag{8}$$

$$I_{f,2} = \frac{V_t}{R_{fc,2} + R_{fw}} = \frac{V_t}{R_{f,2}} = \frac{240}{120} = 2 \ A \tag{9}$$

The equation of the second segment of the magnetization characteristics curve of machine can be formulated as follows:

$$E_a - 180 = \frac{240 - 180}{4 - 1}(I_f - 1) \Rightarrow E_a = 20I_f + 160 \tag{10}$$

By using the magnetization characteristics curve of machine for $I_{f,\ 1} = 3 \ A$ and $I_{f,\ 2} = 2 \ A$, we have:

$$I_{f,1} = 3 \ A \Rightarrow \widehat{E}_{a,1} = 220 \ V \tag{11}$$

$$I_{f,2} = 2 \ A \Rightarrow \widehat{E}_{a,2} = 200 \ V \tag{12}$$

As we know, the induced armature voltage of an electric machine can be determined by using $E_a = k_a \varphi \omega$. Therefore, we can write:

$$\frac{E_{a,2}}{E_{a,1}} = \frac{\varphi_2 \omega_2}{\varphi_1 \omega_1} = \frac{\varphi_2 n_2}{\varphi_1 n_1} \tag{13}$$

Since both induced armature voltages ($\widehat{E}_{a,1}$ and $\widehat{E}_{a,2}$) have been extracted from the magnetization characteristics curve, we have $n_1 = n_2$. Hence:

$$\frac{\widehat{E}_{a,2}}{\widehat{E}_{a,1}} = \frac{\varphi_2}{\varphi_1} = \frac{200}{220} \tag{14}$$

As we know, the electromagnetic torque of an electric machine can be determined by using $T_e = k_a \varphi I_a$. Therefore, we can write:

$$\frac{T_{e,2}}{T_{e,1}} = \frac{\varphi_2 I_{a,2}}{\varphi_1 I_{a,1}} \tag{15}$$

Solving (14) and (15):

$$\frac{T_{e,2}}{T_{e,1}} = \frac{200}{220} \times \frac{I_{a,2}}{I_{a,1}} \tag{16}$$

Solving (1) and (16):

$$\frac{n_2}{n_1} = \frac{200}{220} \times \frac{I_{a,2}}{I_{a,1}} \tag{17}$$

$$\Rightarrow \frac{n_2}{1500} = \frac{200}{220} \times \frac{77}{50} \Rightarrow n_2 = 2100 \; rpm$$

Choice (3) is the answer.

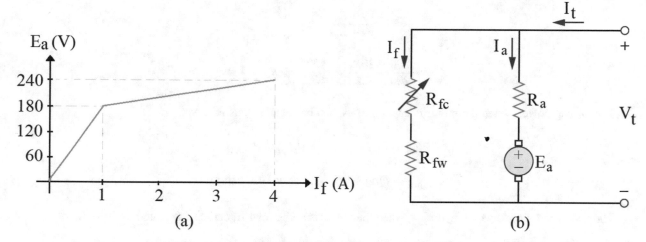

Fig. 16.15 The graph and electrical circuit of solution of problem 16.15

16.16. Based on the information given in the problem, we have:

$$V_t = 200 \; V \tag{1}$$

$$\omega_A = 157 \; rad/sec \tag{2}$$

$$I_{t,A} = 20 \; A \tag{3}$$

$$\omega_B = 167.5 \; rad/sec \tag{4}$$

$$I_{t,B} = 18 \; A \tag{5}$$

$$I_f \approx 0 \; A \tag{6}$$

From (6), we have:

$$I_a = I_t \tag{7}$$

Method 1: As we know, the induced armature voltage of an electric machine can be determined by using $E_a = k_a\varphi\omega$. Therefore, we can write:

$$\frac{E_{a,A}}{E_{a,B}} = \frac{\varphi_A\omega_A}{\varphi_B\omega_B} \tag{8}$$

Since $V_{t, B} = V_{t, A}$, we have:

$$I_{f,B} = I_{f,A} \Rightarrow \varphi_B = \varphi_A \tag{9}$$

Therefore:

$$\frac{E_{a,A}}{E_{a,B}} = \frac{\omega_A}{\omega_B} \tag{10}$$

Figure 16.16 shows the electrical circuit of a shunt DC motor. Applying KVL in the right-hand side mesh of the circuit:

$$E_a = V_t - R_a I_a \tag{11}$$

Solving (10) and (11):

$$\frac{V_t - R_a I_{a,A}}{V_t - R_a I_{a,B}} = \frac{\omega_A}{\omega_B} \tag{12}$$

$$\Rightarrow \frac{200 - 20R_a}{200 - 18R_a} = \frac{157}{167.5} \Rightarrow R_a \simeq 4 \ \Omega \tag{13}$$

The air gap power, when the motor is under load A, can be calculated as follows:

$$P_{ag,A} = P_{in,A} - P_{cu,A} = V_t I_{t,A} - R_a I_{a,A}{}^2 \tag{14}$$

$$\Rightarrow P_{ag,A} = (200 \times 20) - \left(4 \times 20^2\right) = 2400 \ W \tag{15}$$

The developed electromagnetic torque, when the motor is under load A, can be calculated as follows:

$$T_{e,A} = \frac{P_{ag,A}}{\omega_A} = \frac{2400}{157} \approx 15.2 \ N.m \tag{16}$$

The air gap power, when the motor is under load B, can be calculated as follows:

$$P_{ag,B} = P_{in,B} - P_{cu,B} = V_t I_{t,B} - R_a I_{a,B}{}^2 \tag{17}$$

$$\Rightarrow P_{ag,B} = (200 \times 18) - \left(4 \times 18^2\right) = 2304 \ W \tag{18}$$

The developed electromagnetic torque, when the motor is under load B, can be calculated as follows:

$$T_{e,B} = \frac{P_{ag,B}}{\omega_B} = \frac{2304}{167.5} \approx 13.7 \ N.m \tag{19}$$

Method 2: As we know, the induced armature voltage of an electric machine can be determined by using $E_a = k_a \varphi \omega = k'_a \omega$. Applying KVL in the right-hand side mesh of the circuit of Fig. 16.16:

$$V_t = E_a + R_a I_a = k'_a \omega + R_a I_a \tag{20}$$

$$\Rightarrow \begin{cases} V_t = k'_a \times \omega_A + R_a I_{a,A} \\ V_t = k'_a \times \omega_B + R_a I_{a,B} \end{cases} \Rightarrow \begin{cases} 200 = k'_a \times 157 + 20R_a \\ 200 = k'_a \times 167.5 + 18R_a \end{cases} \tag{21}$$

$$\Rightarrow R_a = 4 \ \Omega, \quad k'_a = 0.7625 \tag{22}$$

As we know, the electromagnetic torque of an electric machine can be determined by using $T_e = k_a \varphi I_a = k_a' I_a$. Therefore, we can write:

$$T_{ag,A} = k_a' I_{a,A} = 0.7625 \times 20 \approx 15.2 \; N.m$$

$$T_{ag,B} = k_a' I_{a,B} = 0.7625 \times 18 \approx 13.7 \; N.m$$

Choice (3) is the answer.

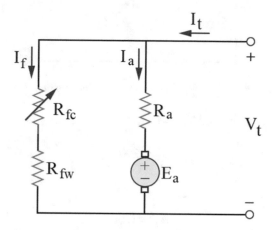

Fig. 16.16 The electrical circuit of solution of problem 16.16

16.17. Based on the information given in the problem, we have:

$$\varphi \propto I_f \tag{1}$$

$$V_{t,2} = 0.5 V_{t,1} \tag{2}$$

As we know, the induced armature voltage of an electric machine can be determined as follows:

$$E_a = k_a \varphi \omega \overset{(1)}{\Longrightarrow} E_a = k_a I_f \omega \tag{3}$$

Figure 16.17 shows the electrical circuit of a shunt DC motor. Applying Ohm's law for the branch of field winding:

$$I_f = \frac{V_t}{R_{fc} + R_{fw}} = \frac{V_t}{R_f} \Rightarrow I_f \propto V_t \tag{4}$$

Solving (3) and (4):

$$E_a = k_a V_t \omega \tag{5}$$

Moreover, as we know, in the no-load condition, we have:

$$E_a \approx V_t \tag{6}$$

Solving (5) and (6):

$$k_a V_t \omega = V_t \Rightarrow \omega = \frac{1}{k_a} = \text{Const.} \Rightarrow \frac{\omega_2}{\omega_1} = 1$$

Choice (3) is the answer.

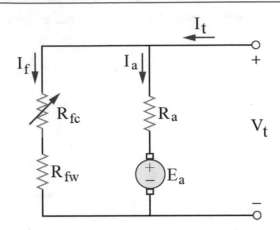

Fig. 16.17 The electrical circuit of solution of problem 16.17

16.18. Based on the information given in the problem, we have:

$$V_t = 200 \ V \tag{1}$$

$$I_t = 20 \ A \tag{2}$$

$$\varphi_2 = 1.1\varphi_1 \tag{3}$$

$$\varphi \propto I_f \tag{4}$$

$$R_a = 0.1 \ \Omega \tag{5}$$

Figure 16.18 shows the electrical circuit of a shunt DC motor. Applying KVL in the right-hand side mesh of the circuit:

$$E_{a,1} = V_t - R_a I_{a,1} = 200 - (0.1 \times 20) = 198 \ V \tag{6}$$

Now, if the magnetic flux suddenly increases to about 10%, the speed of motor can be assumed constant ($\omega_2 = \omega_1$) for a short period of time. The new electromotive force (emf) of the machine can be achieved as follows, since the magnetic circuit of machine is linear:

$$\varphi_2 = 1.1\varphi_1 \xrightarrow{\ E_a \propto \varphi \ } E_{a,2} = 1.1E_{a,1} = 1.1 \times 198 = 217.8 \tag{7}$$

As can be seen, $E_{a,\,2} > V_t$. Therefore, the machine is now behaving like a generator.

Applying KVL in the right-hand side mesh of the circuit:

$$E_{a,2} = V_t - R_a I_{a,2} \tag{8}$$

$$\Rightarrow 217.8 = 200 - (0.1 I_{a_2}) \Rightarrow I_{a,2} = -178 \ A \Rightarrow |I_{a,2}| = 178 \ A$$

The negative sign of the current indicates that the current direction has changed and shows that the machine is now a generator.

Choice (1) is the answer.

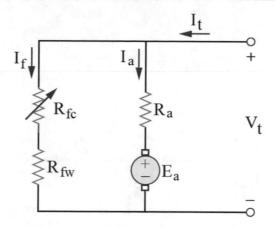

Fig. 16.18 The electrical circuit of solution of problem 16.18

16.19. Based on the information given in the problem, we have:

$$V_{t2} = 0.8V_{t1} \tag{1}$$

$$T_{load} = \text{Const.} \Rightarrow T_e = \text{Const.} \tag{2}$$

$$R_a \approx 0 \ \Omega \tag{3}$$

Since the voltage suddenly decreases, the speed of the motor will remain constant for a short period of time. In other words:

$$\omega_2 = \omega_1 \tag{4}$$

Figure 16.19 shows the electrical circuit of a shunt DC motor. Applying KVL in the right-hand side mesh of the circuit:

$$E_a = V_t - R_a I_a \xrightarrow{R_a \approx 0} E_a = V_t \tag{5}$$

Solving (1) and (5):

$$E_{a,2} = 0.8E_{a,1} \tag{6}$$

As we know, the induced armature voltage of an electric machine can be determined by using $E_a = k_a \varphi \omega$. Therefore, we have:

$$\frac{E_{a,1}}{E_{a,2}} = \frac{\varphi_1 \omega_1}{\varphi_2 \omega_2} \tag{7}$$

Solving (4), (6), and (7):

$$\frac{1}{0.8} = \frac{\varphi_1}{\varphi_2} \tag{8}$$

As we know, the electromagnetic torque of an electric machine can be determined by using $T_e = k_a \varphi I_a$. Therefore, we can write:

$$\frac{T_{e,1}}{T_{e,2}} = \frac{\varphi_1 I_{a,1}}{\varphi_2 I_{a,2}} \tag{9}$$

Solving (2), (8), and (9):

$$1 = \frac{1}{0.8} \times \frac{I_{a,1}}{I_{a,2}} \Rightarrow I_{a,2} = 1.25 I_{a,1}$$

Choice (1) is the answer.

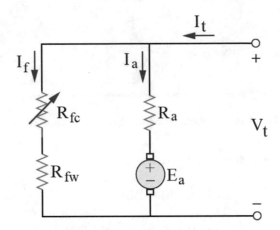

Fig. 16.19 The electrical circuit of solution of problem 16.19

16.20. Based on the information given in the problem, the magnetization characteristics of the machine at the speed of 1800 rpm are as follows:

$$E_a = \frac{500 I_f}{2 + I_f} \tag{1}$$

Moreover, we have:

$$V_{t,1} = 200 \ V \tag{2}$$

$$n_1 = 1800 \ rpm \tag{3}$$

$$I_{a,1} = 100 \ A \tag{4}$$

$$V_{t,2} = 150 \ V \tag{5}$$

$$R_a \approx 0 \ \Omega \tag{6}$$

Figure 16.20 shows the electrical circuit of a shunt DC motor. Applying KVL in the right-hand side mesh of the circuit:

$$E_a = V_t - R_a I_a \xrightarrow{R_a \approx 0} E_a = V_t \tag{7}$$

Solving (1), (2), and (7):

$$200 = \frac{500 I_{f,1}}{2 + I_{f,1}} \Rightarrow I_{f,1} = 2 \ A \tag{8}$$

Applying Ohm's law for the branch of field winding for the first condition:

$$R_f = R_{fc} + R_{fw} = \frac{V_{t,1}}{I_{f,1}} = \frac{200}{2} = 100 \ \Omega \tag{9}$$

Applying Ohm's law for the branch of field winding for the second condition:

$$I_{f,2} = \frac{V_{t,2}}{R_f} = \frac{150}{100} = 1.5\,A \tag{10}$$

The magnetization characteristics of the machine at the new speed can be determined as follows:

$$E_{a,2}|_{n_2} = E_{a,1}|_{n_1} \times \frac{n_2}{n_1} \Rightarrow E_{a,2}|_{n_2} = \frac{500 I_{f,2}}{I_{f,2}+3} \times \frac{n_2}{1800} \tag{11}$$

Solving (11) and considering the value of parameters:

$$\Rightarrow 150 = \frac{500 \times 1.5}{1.5+3} \times \frac{n_2}{1800} \Rightarrow n_2 = 1620\,rpm$$

Choice (3) is the answer.

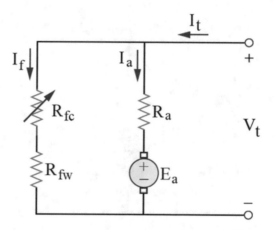

Fig. 16.20   The electrical circuit of solution of problem 16.20

16.21. Based on the information given in the problem, we have:

$$V_t = 260\,V \tag{1}$$

$$T_{load} = \text{Const.} \tag{2}$$

$$R_a = 0.25\,\Omega \tag{3}$$

$$I_{a,1} = 40\,A \tag{4}$$

$$n_1 = 800\,rpm \tag{5}$$

$$\varphi_2 = 0.8\varphi_1 \tag{6}$$

As we know, the electromagnetic torque of an electric machine can be determined by using $T_e = k_a\varphi I_a$. Therefore, we can write:

$$\frac{T_{e,1}}{T_{e,2}} = \frac{\varphi_1 I_{a,1}}{\varphi_2 I_{a,2}} \Rightarrow 1 = \frac{\varphi_1 \times 40}{0.8\varphi_1 \times I_{a,2}} \Rightarrow I_{a,2} = 50\,A \tag{7}$$

Figure 16.21 shows the electrical circuit of a shunt DC motor. Applying KVL in the right-hand side mesh of the circuit:

$$E_{a,1} = V_t - R_a I_{a,1} = 260 - (0.25 \times 40) = 250\ V \tag{8}$$

$$E_{a,2} = V_t - R_a I_{a,2} = 260 - (0.25 \times 50) = 247.5\ V \tag{9}$$

As we know, the induced armature voltage of an electric machine can be determined by using $E_a = k_a \varphi \omega$. Therefore, we have:

$$\frac{E_{a,1}}{E_{a,2}} = \frac{\varphi_1 \omega_1}{\varphi_2 \omega_2} = \frac{\varphi_1 n_1}{\varphi_2 n_2} \tag{10}$$

$$\Rightarrow \frac{250}{247.5} = \frac{\varphi_1 \times 800}{0.8\varphi_1 \times n_2} \Rightarrow n_2 = 990\ rpm$$

Choice (4) is the answer.

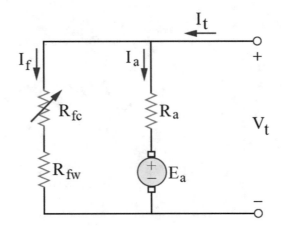

Fig. 16.21 The electrical circuit of solution of problem 16.21

16.22. The magnetization characteristics of a DC shunt motor at a specific speed are shown in Fig. 16.22.a. Moreover, based on the information given in the problem, we have:

$$T_{load} = \text{Const.} \Rightarrow T_e = \text{Const.} \tag{1}$$

$$V_t = 100\ V \tag{2}$$

$$I_{a,1} = 1\ A \tag{3}$$

$$R_a = 0.5\ \Omega \tag{4}$$

$$R_f = 100\ \Omega \tag{5}$$

$$n_1 = 500\ rpm \tag{6}$$

$$I_{f,2} = 0\ A \tag{7}$$

Figure 16.22.b shows the electrical circuit of a shunt DC machine. Applying Ohm's law for the branch of field winding:

$$I_{f,1} = \frac{V_t}{R_{fc} + R_{fw}} = \frac{V_t}{R_f} = \frac{100}{100} = 1 \, A \tag{8}$$

The primary electromotive force (emf) of the machine can be achieved from its magnetization characteristics as follows:

$$I_{f,1} = 1 \, A \Rightarrow \widehat{E}_{a,1} = 200 \, V \tag{9}$$

Likewise, the secondary emf of the machine can be determined as follows:

$$I_{f,2} = 0 \Rightarrow \widehat{E}_{a,2} = 20 \, V \tag{10}$$

As we know, the emf of an electric machine can be determined by using $E_a = k_a \varphi \omega$. Therefore, we have:

$$\frac{E_{a,1}}{E_{a,2}} = \frac{\varphi_1 \omega_1}{\varphi_2 \omega_2} = \frac{\varphi_1 n_1}{\varphi_2 n_2} \tag{11}$$

Since the field current is suddenly interrupted, the speed of the motor can be assumed constant ($n_2 = n_1$) for a short period of time. Therefore:

$$\frac{\widehat{E}_{a,1}}{\widehat{E}_{a,2}} = \frac{\varphi_1}{\varphi_2} \Rightarrow \frac{\varphi_1}{\varphi_2} = \frac{200}{20} = 10 \tag{12}$$

As we know, the electromagnetic torque of an electric machine can be determined by using $T_e = k_a \varphi I_a$. Therefore, we can write:

$$\frac{T_{e,1}}{T_{e,2}} = \frac{\varphi_1 I_{a,1}}{\varphi_2 I_{a,2}} \tag{13}$$

Solving (13) and considering the value of parameters:

$$1 = 10 \times \frac{1}{I_{a,2}} \Rightarrow I_{a,2} = 10 \, A \tag{14}$$

Applying KVL in the right-hand side mesh of the circuit for the first case:

$$E_{a,1} = V_t - R_a I_{a,1} = 100 - (0.5 \times 1) = 99.5 \, V \tag{15}$$

Applying KVL in the right-hand side mesh of the circuit for the second case:

$$E_{a,2} = V_t - R_a I_{a,2} = 100 - (0.5 \times 10) = 95 \, V \tag{16}$$

Solving (11) and considering the value of parameters:

$$\frac{99.5}{95} = 10 \times \frac{500}{n_2} \Rightarrow n_2 = 4750 \, rpm$$

Choice (4) is the answer.

As can be noticed, by interrupting the field current, and consequently by the remarkable reduction of the magnetic flux of DC shunt motor, the speed of motor aggressively increases ($n_1 = 500 \, rpm \rightarrow n_2 = 4750 \, rpm$), and the armature winding current significantly increases ($I_{a, \, 1} = 1 \, A \rightarrow I_{a, \, 2} = 10 \, A$) that can damage the insulations of armature winding.

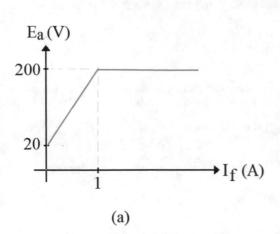

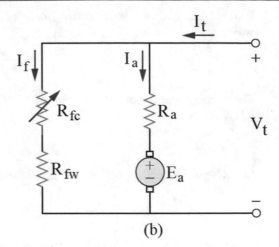

(a)   (b)

Fig. 16.22 The graph and electrical circuit of solution of problem 16.22

16.23. Based on the information given in the problem, we have:

$$P = 40\ kW \tag{1}$$

$$V_t = 500\ V \tag{2}$$

$$R_a = 0.1\ \Omega \tag{3}$$

$$R_f = 250\ \Omega \tag{4}$$

$$n_{NL} = 1500\ rpm \tag{5}$$

$$I_{t,NL} = 12\ A \tag{6}$$

$$I_{t,FL} = 70\ A \tag{7}$$

$$\varphi_{FL} = 0.95\varphi_{NL} \tag{8}$$

Figure 16.23 shows the electrical circuit of a shunt DC motor. Applying Ohm's law for the branch of field winding:

$$I_{f,NL} = \frac{V_t}{R_{fc} + R_{fw}} = \frac{V_t}{R_f} = \frac{500}{200} = 2\ A \tag{9}$$

Since $V_{t,FL} = V_{t,NL}$, we have:

$$I_{f,FL} = I_{f,NL} = 2\ A \tag{10}$$

Applying KCL at the top node of the circuit:

$$I_{a,NL} = I_{t,NL} - I_{f,NL} = 12 - 2 = 10\ A \tag{11}$$

Applying KVL in the right-hand side mesh of the circuit:

$$E_{a,NL} = V_t - R_a I_{a,NL} = 500 - (0.1 \times 10) = 499\ V \tag{12}$$

Applying KCL at the top node of the circuit:

$$I_{a,FL} = I_{t,FL} - I_{f,FL} = 70 - 2 = 68\ A \tag{13}$$

Applying KVL in the right-hand side mesh of the circuit:

$$E_{a,FL} = V_t - R_a I_{a,FL} = 500 - (0.1 \times 68) = 493.2 \ V \tag{14}$$

The developed electromagnetic torque in the full-load condition can be calculated as follows:

$$T_{e,NL} = \frac{P_{e,NL}}{\omega_{NL}} = \frac{E_{a,NL} I_{a,NL}}{n_{NL} \frac{2\pi}{60}} = \frac{499 \times 10}{1500 \times \frac{2\pi}{60}} = 31.77 \ N.m \tag{15}$$

As we know, the electromagnetic torque of an electric machine can be determined by using $T_e = k_a \varphi I_a$. Therefore, we can write:

$$\frac{T_{e,NL}}{T_{e,FL}} = \frac{\varphi_{FL} I_{a,NL}}{\varphi_{NL} I_{a,FL}} \tag{16}$$

Solving (8) and (16):

$$\frac{T_{e,NL}}{T_{e,FL}} = \frac{\varphi_{NL}}{\varphi_{FL}} \times \frac{I_{a,NL}}{I_{a,FL}} \tag{17}$$

$$\Rightarrow \frac{31.77}{T_{e,FL}} = \frac{1}{0.95} \times \frac{10}{68} \Rightarrow T_{e,FL} \approx 205 \ N.m$$

As we know, the induced armature voltage of an electric machine can be determined by using $E_a = k_a \varphi \omega_m$. Therefore, we can write:

$$\frac{E_{a,NL}}{E_{a,FL}} = \frac{\varphi_{NL} \omega_{NL}}{\varphi_{FL} \omega_{FL}} = \frac{\varphi_{NL} n_{NL}}{\varphi_{FL} n_{FL}} \tag{18}$$

Solving (8) and (17):

$$\frac{E_{a,NL}}{E_{a,FL}} = \frac{n_{NL}}{0.95 \times n_{FL}} \tag{19}$$

$$\Rightarrow \frac{499}{493.2} = \frac{1500}{0.95 \times n_{FL}} \Rightarrow n_{FL} \approx 1561 \ rpm$$

Choice (4) is the answer.

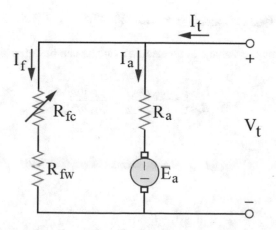

Fig. 16.23 The electrical circuit of solution of problem 16.23

16.24. Based on the information given in the problem, we have:

$$V_t = 220 \ V \tag{1}$$

$$R_a = 0.1 \ \Omega \tag{2}$$

$$R_f = 100 \ \Omega \tag{3}$$

$$I_{t,NL} = 5 \ A \tag{4}$$

$$n_{NL} = 1000 \ rpm \tag{5}$$

$$I_{t,FL} = 50 \ A \tag{6}$$

The electrical circuit of a shunt DC motor is shown in Fig. 16.24. Applying KCL at the top node in the no-load condition:

$$I_{a,NL} = I_{t,NL} - I_{f,NL} = I_{t,NL} - \frac{V_t}{R_{fc} + R_{fw}} = I_{t,NL} - \frac{V_t}{R_f} = 5 - \frac{220}{100} = 2.8 \ A \tag{7}$$

Applying KVL in the right-hand side mesh of the circuit:

$$E_{a,NL} = V_t - R_a I_{a,NL} = 220 - (0.1 \times 2.8) = 219.72 \ V \tag{8}$$

The rotational power loss (P_{rot}) in the no-load condition is equal to the air gap power (P_{ag}) that can be calculated as follows:

$$P_{rot} = P_{ag,NL} = E_{a,NL} I_{a,NL} = 219.72 \times 2.8 = 615.2 \ W \tag{9}$$

Applying KCL at the top node in the full-load condition:

$$I_{a,FL} = I_{t,FL} - I_{f,FL} = I_{t,FL} - \frac{V_t}{R_{fc} + R_{fw}} = I_{t,FL} - \frac{V_t}{R_f} = 50 - \frac{220}{100} = 47.8 \ A \tag{10}$$

Applying KVL in the right-hand side mesh of the circuit:

$$E_{a,FL} = V_t - R_a I_{a,FL} = 220 - (0.1 \times 47.8) = 215.22 \ V \tag{11}$$

The air gap power in the full-load condition can be calculated as follows:

$$P_{ag,FL} = E_{a,FL} I_{a,FL} = 215.22 \times 47.8 = 10287.5 \ W \tag{12}$$

As we know, the induced armature voltage of an electric machine can be determined by using $E_a = k_a \varphi \omega$. Therefore, we can write:

$$\frac{E_{a,NL}}{E_{a,FL}} = \frac{\varphi_{NL} \omega_{NL}}{\varphi_{FL} \omega_{FL}} = \frac{\varphi_{NL} n_{NL}}{\varphi_{FL} n_{FL}} \tag{13}$$

Since $V_{t,NL} = V_{t,FL}$, we have:

$$I_{f,NL} = I_{f,FL} \Rightarrow \varphi_{NL} = \varphi_{FL} \tag{14}$$

Therefore:

$$\frac{E_{a,NL}}{E_{a,FL}} = \frac{n_{NL}}{n_{FL}} \Rightarrow \frac{219.72}{215.22} = \frac{1000}{n_{FL}} \Rightarrow n_{FL} = 979.5 \ rpm \tag{15}$$

The output electromagnetic torque in the full-load condition can be calculated as follows:

$$T_{out,FL} = \frac{P_{out,FL}}{\omega_{FL}} = \frac{P_{out,FL}}{n_{FL}\frac{2\pi}{60}} = \frac{P_{ag,FL} - P_{rot}}{n_{FL}\frac{2\pi}{60}} \tag{16}$$

$$\Rightarrow T_{out,FL} = \frac{10287.5 - 615.2}{2\pi \times \frac{979.5}{60}} \Rightarrow T_{out,FL} \approx 94 \, N.m$$

Choice (2) is the answer.

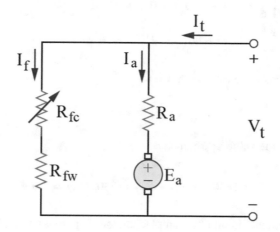

Fig. 16.24 The electrical circuit of solution of problem 16.24

16.25. Based on the information given in the problem, the electromotive force (emf) of the machine at the speed of 1200 rpm is as follows:

$$E_{a,1} = \begin{cases} 10 + 115I_f, & 0 \le I_f \le 2A \tag{1} \\ 240 + 80(I_f - 2), & I_f \ge 2A \tag{2} \end{cases}$$

In addition, we have:

$$n_1 = 1000 \, rpm \tag{3}$$

$$R_{fw} = 50 \, \Omega \tag{4}$$

$$R_a \approx 0 \, \Omega \tag{5}$$

$$V_t \approx 240 \, V \tag{6}$$

$$n_2 = 1000 \, rpm \tag{7}$$

As we know, the induced armature voltage of an electric machine can be determined by using $E_a = k_a\varphi\omega_m$. Therefore, we can write:

$$\frac{E_{a,2}}{E_{a,1}} = \frac{\varphi_2\omega_2}{\varphi_1\omega_1} = \frac{\varphi_2 n_2}{\varphi_1 n_1} \tag{8}$$

Since the terminal voltage is constant, we have:

$$I_{f,2} = I_{f,1} \Rightarrow \varphi_2 = \varphi_1 \tag{9}$$

Therefore:

$$\frac{E_{a,2}}{E_{a,1}} = \frac{n_2}{n_1} \tag{10}$$

The emf of the machine at the speed of 1000 rpm can be updated as follows:

$$E_{a,2} = \frac{1000}{1200} \times E_{a,1}$$

$$= \begin{cases} \frac{5}{6} \times \left(10 + 115 I_f\right), & 0 \le I_f \le 2A \tag{11} \\ \frac{5}{6} \times \left(240 + 80\left(I_f - 2\right)\right), & I_f \ge 2A \tag{12} \end{cases}$$

Since $R_a \approx 0\ \Omega$, in the no-load condition, we have:

$$E_a \approx V_t \tag{13}$$

Solving (12) by considering the value of parameters:

$$\frac{5}{6} \times \left(240 + 80\left(I_f - 2\right)\right) = 240 \Rightarrow I_f = 2.6\,A \tag{14}$$

Figure 16.25 shows the electrical circuit of a shunt DC motor. Applying Ohm's law for the branch of field winding:

$$R_{fw} + R_{fc} = \frac{V_t}{I_f} \Rightarrow 50 + R_{fc} = \frac{240}{2.6} \Rightarrow R_{fc} \approx 42\ \Omega$$

Choice (3) is the answer.

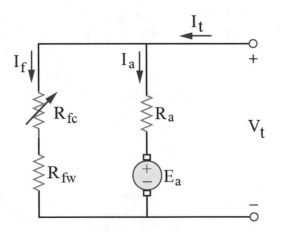

Fig. 16.25 The electrical circuit of solution of problem 16.25

16.26. Based on the information given in the problem, the magnetization characteristics of the DC shunt motor at a specific speed are illustrated in Fig. 16.26.a. In addition, we have:

$$V_t = 300\ V \tag{1}$$

$$n_1 = 1200\ rpm \tag{2}$$

$$R_{f,1} = 100 \ \Omega \tag{3}$$

$$R_{a,1} \approx 0 \ \Omega \tag{4}$$

$$R_{f,2} = 150 \ \Omega \tag{5}$$

$$R_{a,1} \approx 1 \ \Omega \tag{6}$$

$$I_{a,2} = 20 \ A \tag{7}$$

Figure 16.26.b shows the electrical circuit of a shunt DC motor. Applying Ohm's law for the branch of field winding for the first condition:

$$I_{f,1} = \frac{V_t}{R_{f,1}} = \frac{300}{100} = 3 \ A \tag{8}$$

The primary electromotive force (emf) of machine can be obtained from the magnetization characteristics of the machine as follows:

$$I_{f,1} = 3 \ A \Rightarrow \widehat{E}_{a,1} = 220 \ V \tag{9}$$

Applying Ohm's law for the branch of field winding for the second condition:

$$I_{f,2} = \frac{V_t}{R_{f,2}} = \frac{300}{150} = 2 \ A \tag{10}$$

Likewise, the secondary emf of machine can be calculated as follows:

$$I_{f,2} = 2A \Rightarrow \widehat{E}_{a,2} = 210 \ V \tag{11}$$

As we know, the emf of an electric machine can be determined by using $E_a = k_a \varphi \omega$. Therefore, we have:

$$\frac{E_{a,1}}{E_{a,2}} = \frac{\varphi_1 \omega_1}{\varphi_2 \omega_2} = \frac{\varphi_1 n_1}{\varphi_2 n_2} \tag{12}$$

The speed of the motor can be assumed constant ($n_2 = n_1$) for a short period of time. Therefore:

$$\frac{\widehat{E}_{a,1}}{\widehat{E}_{a,2}} = \frac{\varphi_1}{\varphi_2} \Rightarrow \frac{\varphi_1}{\varphi_2} = \frac{220}{210} \tag{13}$$

Applying KVL in the right-hand side mesh of the circuit (the first condition):

$$E_{a,1} = V_t - R_{a,1} I_{a,1} \xrightarrow{R_{a,1} \approx 0} E_{a,1} = 300 \ V \tag{14}$$

Applying KVL in the right-hand side mesh of the circuit (the second condition):

$$E_{a,2} = V_t - R_a I_{a,2} = 300 - (1 \times 20) = 280 \ V \tag{15}$$

Solving (12)–(15):

$$\frac{300}{280} = \frac{220}{210} \times \frac{1200}{n_2} \Rightarrow n_2 = 1173.3 \; rpm$$

Choice (1) is the answer.

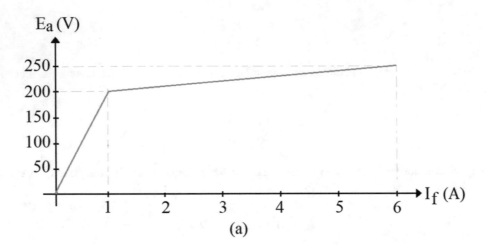

(a)

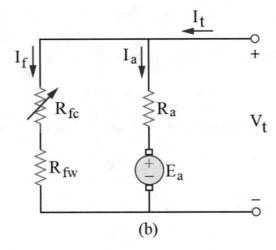

(b)

Fig. 16.26 The graph and electrical circuit of solution of problem 16.26

Abstract

In this chapter, the problems related to the series DC electric motors are solved. In this chapter, the problems are categorized in different levels based on their difficulty levels (easy, normal, and hard) and calculation amounts (small, normal, and large). Additionally, the problems are ordered from the easiest problem with the smallest computations to the most difficult problems with the largest calculations.

17.1. The speed of a 220 V series DC motor is adjusted by a field control rheostat and its armature winding resistance is 0.1 Ω. When the rheostat is out of the circuit, the armature current is 20 A and the speed of motor is 1000 rpm. What is the resistance of rheostat when the armature current is 16 A and the motor is rotating at the speed of 1206 rpm?

Difficulty level   ○ Easy   ● Normal   ○ Hard
Calculation amount   ○ Small   ● Normal   ○ Large

1) 0.1 Ω
2) 0.5 Ω
3) 0.2 Ω
4) 0.3 Ω

17.2. A 300V series DC motor has the armature winding resistance of 0.1 Ω, is rotating with the speed of 1200 rpm, and is drawing the current of 50 A. If the maximum speed of motor is 3000 rpm, calculate its minimum current assuming that the magnetic circuit of motor is linear.

Difficulty level   ○ Easy   ● Normal   ○ Hard
Calculation amount   ○ Small   ● Normal   ○ Large

1) 25 A
2) 50 A
3) 35 A
4) 20 A

17.3. A series DC motor is drawing 30 A in the saturation condition. In this condition, the developed electromagnetic torque of motor is proportional to which one of the following numbers?

Difficulty level   ○ Easy   ○ Normal   ● Hard
Calculation amount   ● Small   ○ Normal   ○ Large

1) 900
2) $\frac{1}{30}$
3) 30
4) $\frac{1}{900}$

© The Author(s), under exclusive license to Springer Nature Switzerland AG 2022
M. Rahmani-Andebili, *DC Electric Machines, Electromechanical Energy Conversion Principles, and Magnetic Circuit Analysis*,
https://doi.org/10.1007/978-3-031-08863-6_17

17.4. A series DC motor is drawing 80 A in the saturation condition. In this condition, calculate the armature current of motor if its load increases to about 10%.

Difficulty level   ○ Easy   ○ Normal   ● Hard
Calculation amount   ● Small   ○ Normal   ○ Large
1) 82.5 A
2) 98 A
3) 88 A
4) 108 A

17.5. One of the speed control methods in a series DC motor is changing the number of turns of the field winding. How the motor speed will change if the number of turns of the field winding decreases from 100 to 81, while the load torque is constant? Herein, assume that the magnetic circuit of motor is linear and the resistance of field and armature windings is negligible.

Difficulty level   ○ Easy   ○ Normal   ● Hard
Calculation amount   ○ Small   ● Normal   ○ Large
1) It will increase about 10%.
2) It will increase about 20%.
3) It will decrease about 10%.
4) It will decrease about 20%.

17.6. A 40 V series DC motor draws the current of 50 A at the speed of 1500 rpm. If its circuit resistance is 0.2 Ω, what resistance needs to be added to the circuit to provide the rated torque at the speed of 1200 rpm? Assume linear magnetization characteristics for the machine.

Difficulty level   ○ Easy   ○ Normal   ● Hard
Calculation amount   ○ Small   ● Normal   ○ Large
1) 0.92 Ω
2) 1.92 Ω
3) 1.73 Ω
4) 0.73 Ω

17.7. If the developed torque of a series DC motor is doubled, how its copper loss will change? Assume linear magnetization characteristics for the machine.

Difficulty level   ○ Easy   ○ Normal   ● Hard
Calculation amount   ○ Small   ● Normal   ○ Large
1) It will increase about 41%.
2) It will increase about 100%.
3) It will decrease about 50%.
4) It will decrease about 30%.

17.8. In a series DC motor, if the number of turns of the series winding is halved, what will be the ratio of the new speed of motor to the old one? Herein, ignore the armature winding resistance and assume a linear magnetic circuit and constant load for the motor.

Difficulty level   ○ Easy   ○ Normal   ● Hard
Calculation amount   ○ Small   ● Normal   ○ Large
1) 2
2) $\frac{1}{2}$
3) $\sqrt{2}$
4) $\frac{1}{\sqrt{2}}$

17.9. In a 200 V series DC motor that draws the current of 50 A and rotates at the speed of 1000 rpm, the armature winding resistance is 0.1 Ω. What external resistance must be added to the armature circuit to decrease the speed of motor to about 900 rpm? Herein, the load torque remains constant.

Difficulty level   ○ Easy   ○ Normal   ● Hard
Calculation amount   ○ Small   ● Normal   ○ Large

1) 0.1 Ω
2) 0.25 Ω
3) 0.4 Ω
4) 0.5 Ω

17.10. In a 400 V series DC motor that draws the full-load current of 50 A and rotates at the speed of 1000 rpm, the armature winding resistance is 0.2 Ω. What is the speed of motor when its load is doubled? Herein, assume a linear magnetic circuit for the motor.

Difficulty level   ○ Easy   ○ Normal   ● Hard
Calculation amount   ○ Small   ● Normal   ○ Large

1) 800 *rpm*
2) 900 *rpm*
3) 600 *rpm*
4) 700 *rpm*

17.11. A series DC motor has a linear magnetic circuit and its field winding resistance is 0.06 Ω. When the motor is under a specific load, its steady-state current is 20 A. If a 0.12 Ω resistor is paralleled with the field circuit and the load is doubled, what will be the motor current in the steady-state condition?

Difficulty level   ○ Easy   ○ Normal   ● Hard
Calculation amount   ○ Small   ● Normal   ○ Large

1) 14.1 A
2) 23.1 A
3) 28.3 A
4) 34.6 A

17.12. A series DC motor has a linear magnetic circuit and its field winding resistance is R_f. If the resistor of $2R_f$ is paralleled with the field circuit and the load is halved, what will be the ratio of the new current of motor to the previous one?

Difficulty level   ○ Easy   ○ Normal   ● Hard
Calculation amount   ○ Small   ● Normal   ○ Large

1) $\sqrt{\frac{2}{3}}$

2) $\frac{2}{\sqrt{3}}$

3) $\sqrt{\frac{3}{2}}$

4) $\frac{\sqrt{3}}{2}$

17.13. If the load of a series DC motor is doubled, how its copper loss will change?

Difficulty level   ○ Easy   ○ Normal   ● Hard
Calculation amount   ○ Small   ● Normal   ○ Large

1) It will be doubled.
2) It will be halved.
3) It will be quadrupled.
4) There will be no change.

17.14. A series DC motor has the magnetization characteristics shown in Fig. 17.1. When the motor is working at full-load, its current and torque are about 40 A and 24 N.m., respectively. If the starting current of motor is 100 A, calculate its starting torque. Herein, ignore the armature reaction.

Difficulty level   ○ Easy   ○ Normal   ● Hard
Calculation amount   ○ Small   ● Normal   ○ Large

1) 60 A
2) 67 A
3) 100 A
4) 150 A

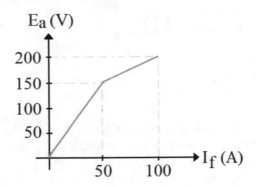

Fig. 17.1 The graph of problem 17.14

17.15. A series DC motor has a linear magnetic circuit. This motor is supplied by a specific voltage and rotates with the speed of n_0. If the number of turns of the field winding and the voltage level are halved and the load torque is doubled, calculate the new speed of the motor. Herein, ignore the ohmic power losses.

Difficulty level	○ Easy	○ Normal	● Hard
Calculation amount	○ Small	● Normal	○ Large

1) $\frac{1}{4}n_0$
2) $\frac{1}{2}n_0$
3) $2n_0$
4) $4n_0$

17.16. A series DC motor has a linear magnetic circuit, is supplied by a 125 V power source, rotates at the speed of 1200 rpm, and draws the current of 50 A. If the voltage level is increased to 190 V and the load torque becomes $2\sqrt{2}$ times, calculate the new speed of motor. Herein, ignore the armature reaction, assume that the armature circuit resistance is 0.1 Ω, and consider that $\varphi \propto \sqrt{I_a}$.

Difficulty level	○ Easy	○ Normal	● Hard
Calculation amount	○ Small	● Normal	○ Large

1) 900 *rpm*
2) $900\sqrt{2}$ *rpm*
3) 1800 *rpm*
4) $1800\sqrt{2}$ *rpm*

17.17. The magnetization characteristics of a series DC motor are shown in Fig. 17.2. The number of turns of the field winding is 25 turns per pole and the total resistance of the armature and field circuits is 2 Ω. The motor is started with a power supply with the voltage of 128 V, and after reaching its final speed, it draws 12 A current. Calculate the ratio of starting torque of motor to its working torque.

Difficulty level	○ Easy	○ Normal	● Hard
Calculation amount	○ Small	○ Normal	● Large

1) 28.44
2) 15.32
3) 20.35
4) 10.67

Fig. 17.2 The graph of problem 17.17

Solutions of Problems: Series DC Electric Motor

18

Abstract

In this chapter, the problems of the 17th chapter are fully solved, in detail, step-by-step, and with different methods.

18.1 Based on the information given in the problem, we have:

$$V_t = 220 \ V \tag{1}$$

$$R_a = 0.1 \ \Omega \tag{2}$$

$$I_{a,1}|_{R_{fc}=0} = 20 \ A \tag{3}$$

$$n_1 = 1000 \ rpm \tag{4}$$

$$I_{a,2} = 16 \ A \tag{5}$$

$$n_2 = 1206 \ rpm \tag{6}$$

$$R_s = 0 \ \Omega \tag{7}$$

Figure 18.1 shows the electrical circuit of a series DC motor. By applying KVL in the loop when the rheostat (field control resistor) is out of the circuit, we have:

$$E_{a,1} = V_t - (R_a + R_s)I_{a,1} = 220 - (0.1 + 0) \times 20 = 218 \ V \tag{8}$$

Applying KVL in the loop when the rheostat (field control resistor) is in the circuit:

$$E_{a,2} = V_t - (R_a + R_s + R_{fc})I_{a,2} = 220 - (0.1 + 0 + R_{fc}) \times 16 = 220 - (0.1 + R_{fc}) \times 16 \tag{9}$$

As we know, the induced armature voltage of an electric machine can be determined by using $E_a = k_a \varphi \omega$. In a series machine with linear magnetization characteristics, we know that $\varphi \propto I_a$. Therefore, we can write:

$$\frac{E_{a,1}}{E_{a,2}} = \frac{\varphi_1 \omega_1}{\varphi_2 \omega_2} = \frac{I_{a,1} n_1}{I_{a,2} n_2} \tag{10}$$

$$\Rightarrow \frac{218}{220 - (0.1 + R_{fc}) \times 16} = \frac{20}{16} \times \frac{1000}{1206} \Rightarrow R_{fc} = 0.5 \ \Omega$$

Choice (2) is the answer.

© The Author(s), under exclusive license to Springer Nature Switzerland AG 2022
M. Rahmani-Andebili, *DC Electric Machines, Electromechanical Energy Conversion Principles, and Magnetic Circuit Analysis*,
https://doi.org/10.1007/978-3-031-08863-6_18

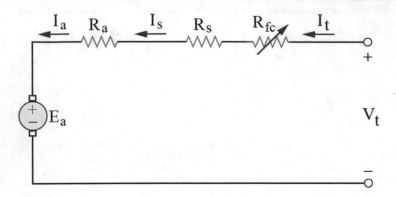

Fig. 18.1 The electrical circuit of solution of problem 18.1

18.2 Based on the information given in the problem, we have:

$$V_t = 300 \ V \tag{1}$$

$$R_a = 0.1 \ \Omega \tag{2}$$

$$n = 1200 \ rpm \tag{3}$$

$$I_t = 50 \ A \tag{4}$$

$$n_{max} = 3000 \ rpm \tag{5}$$

$$\varphi \propto I_a \tag{6}$$

$$R_s = 0 \ \Omega \tag{7}$$

$$R_{fc} = 0 \ \Omega \tag{8}$$

Figure 18.2 shows the electrical circuit of a series DC motor. Applying KVL in the loop (the first condition):

$$E_{a,1} = V_t - \left(R_a + R_s + R_{fc}\right)I_{a,1} = 300 - (0.1 + 0 + 0) \times 50 = 295 \ V \tag{9}$$

Applying KVL in the loop (the second condition):

$$E_{a,2} = V_t - \left(R_a + R_s + R_{fc}\right)I_{a,2} = 300 - (0.1 + 0 + 0)I_{a,2} = 300 - 0.1I_{a,2} \tag{10}$$

As we know, the induced armature voltage of an electric machine can be determined by using $E_a = k_a\varphi\omega$. In a series machine with linear magnetization characteristics, we have $\varphi \propto I_a$. Therefore, we can write:

$$\frac{E_{a,1}}{E_{a,2}} = \frac{\varphi_1\omega_1}{\varphi_2\omega_2} = \frac{I_{a,1}n_1}{I_{a,2}n_2} \tag{11}$$

$$\Rightarrow \frac{295}{300 - 0.1I_{a,2}} = \frac{50 \times 1200}{I_{a,2} \times 3000} \Rightarrow I_{a,2} \approx 20 \ A$$

Choice (4) is the answer.

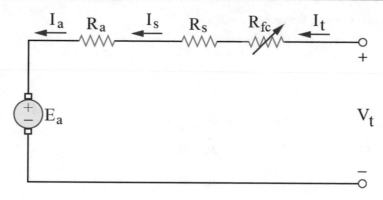

Fig. 18.2 The electrical circuit of solution of problem 18.2

18.3 Based on the information given in the problem, we have:

$$I_{a,sat} = 30\,A \tag{1}$$

As we know, in saturation condition, the developed electromagnetic torque of machine can be determined by using the relation below:

$$T_{e,sat} = k'_a I_{a,sat} \tag{2}$$

Solving (1) and (2):

$$T_{e,sat} = 30k'_a \Rightarrow T_{e,sat} \propto 30$$

Among the choices, only the third choice can be an answer. Choice (3) is the answer.

18.4 Based on the information given in the problem, we have:

$$I_{t,sat} = 80\,A \tag{1}$$

$$T_{load,2} = 1.1 T_{load,1} \Rightarrow T_{e,sat,2} = 1.1 T_{e,sat,1} \tag{2}$$

In the saturation condition, the developed electromagnetic torque of machine can be determined as follows:

$$T_{e,sat} = k'_a I_{a,sat} \tag{3}$$

Therefore:

$$\frac{T_{e,sat,1}}{T_{e,sat,2}} = \frac{I_{a,sat,1}}{I_{a,sat,2}} \tag{4}$$

$$\Rightarrow \frac{1}{1.1} = \frac{I_{a,sat,1}}{I_{a,sat,2}} \Rightarrow I_{a,sat,2} = 1.1 I_{a,sat,1} = 1.1 \times 80 \Rightarrow I_{a,sat,2} = 88\,A$$

Choice (3) is the answer.

18.5 Based on the information given in the problem, we have:

$$N_{s,1} = 100 \text{ turns} \tag{1}$$

$$N_{s,2} = 81 \text{ turns} \tag{2}$$

$$T_{load} = \text{Const.} \tag{3}$$

$$R_s = 0 \ \Omega \tag{4}$$

$$R_a = 0 \ \Omega \tag{5}$$

$$R_{fc} = 0 \ \Omega \tag{6}$$

Since the magnetic circuit of machine is linear, we have:

$$\varphi \propto N_s I_s \tag{7}$$

From (3), we can conclude that the developed electromagnetic torque is constant:

$$T_e = \text{Const.} \tag{8}$$

As we know, the electromagnetic torque of an electric machine can be determined by using $T = k_a \varphi I_a$. Therefore, we can write:

$$\Rightarrow k_a \varphi_1 I_{a,1} = k_a \varphi_2 I_{a,2} \Rightarrow \varphi_1 I_{a,1} = \varphi_2 I_{a,2} \tag{9}$$

Solving (7) and (9) and considering the fact that in a series machine we have $I_s = I_a$:

$$(N_{s,1} I_{a,1}) I_{a,1} = (N_{s,2} I_{a,2}) I_{a,2} \Rightarrow N_{s,1} I_{a,1}^2 = N_{s,2} I_{a,2}^2$$

$$\Rightarrow 100 I_{a,1}^2 = 81 I_{a,2}^2 \Rightarrow I_{a,1} = 0.9 I_{a,2} \tag{10}$$

Figure 18.3 shows the electrical circuit of a series DC motor. By applying KVL in the loop, we have:

$$E_{a,1} = V_t - (R_a + R_s + R_{fc}) I_{a,1} = V_t - 0 = V_t \tag{11}$$

$$E_{a,2} = V_t - (R_a + R_s + R_{fc}) I_{a,2} = V_t - 0 = V_t \tag{12}$$

As we know, the induced armature voltage of an electric machine can be determined by using $E_a = k_a \varphi \omega$. Therefore, we have:

$$\frac{E_{a,1}}{E_{a,2}} = \frac{\varphi_1 \omega_1}{\varphi_2 \omega_2} \stackrel{(7)}{\Longrightarrow} \frac{E_{a,1}}{E_{a,2}} = \frac{N_{s,1} I_{a,1} n_1}{N_{s,2} I_{a,2} n_2} \tag{13}$$

$$\Rightarrow \frac{V_t}{V_t} = \frac{N_{s,1} I_{a,1} n_1}{N_{s,2} I_{a,2} n_2} \Rightarrow 1 = \frac{100 \times 0.9 I_{a,2} n_1}{81 \times I_{a,2} n_2} \Rightarrow \frac{n_2}{n_1} = \frac{90}{81} \Rightarrow n_2 = 1.1 n_1$$

Therefore, the speed of motor will increase about 10%. Choice (1) is the answer.

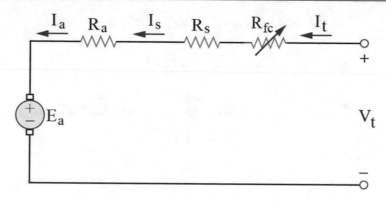

Fig. 18.3 The electrical circuit of solution of problem 18.5

18.6 Based on the information given in the problem, we have:

$$V_t = 40 \ V \tag{1}$$

$$I_t = 50 \ A \tag{2}$$

$$n_1 = 1500 \ rpm \tag{3}$$

$$R_a + R_s = 0.2 \ \Omega \tag{4}$$

$$R_{fc,1} = 0 \ \Omega \tag{5}$$

$$n_2 = 1200 \ rpm \tag{6}$$

$$T_{e,1} = T_{e,2} \tag{7}$$

As we know, the electromagnetic torque of an electric machine can be determined by using $T = k_a \varphi I_a$. In a series machine with linear magnetization characteristics, we know that $\varphi \propto I_a$ and $I_a = I_t$. Therefore, we can write:

$$\frac{T_{e,1}}{T_{e,2}} = \left(\frac{I_{a,1}}{I_{a,2}}\right)^2 \Rightarrow 1 = \left(\frac{50}{I_{a,2}}\right)^2 \Rightarrow I_{a,2} = 50 \ A \tag{8}$$

Figure 18.4 shows the electrical circuit of a series DC motor. Applying KVL in the loop for the first condition:

$$E_{a,1} = V_t - \left(R_a + R_s + R_{fc,1}\right)I_{a,1} = 240 - 50 \times (0.2 + 0 + 0) = 230 \ V \tag{9}$$

As we know, the induced armature voltage of an electric machine can be determined by using $E_a = k_a \varphi \omega$. In a series machine with linear magnetization characteristics, we have $\varphi \propto I_a$. Therefore, we can write:

$$\frac{E_{a,1}}{E_{a,2}} = \frac{\varphi_1 \omega_1}{\varphi_2 \omega_2} = \frac{I_{a,1} n_1}{I_{a,2} n_2} \tag{10}$$

$$\Rightarrow \frac{230}{E_{a,2}} = \frac{1500}{1200} \times \frac{50}{50} \Rightarrow E_{a,2} = 184 \ V \tag{11}$$

Applying KVL in the loop for the second condition:

$$E_{a,2} = V_t - \left(R_a + R_s + R_{fc,2}\right)I_{a,2} \tag{12}$$

$$\Rightarrow 184 = 240 - \left(0.2 + 0 + R_{fc,2}\right) \times 50 \Rightarrow R_{fc,2} = 0.92 \ \Omega$$

Choice (1) is the answer.

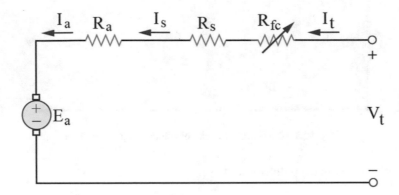

Fig. 18.4 The electrical circuit of solution of problem 18.6

18.7 Based on the information given in the problem, we have:

$$T_{e,2} = 2T_{e,1} \tag{1}$$

As we know, the electromagnetic torque of an electric machine can be determined by using $T = k_a \varphi I_a$. In a series machine with linear magnetization characteristics, we know that $\varphi \propto I_a$ and $I_a = I_t$. Therefore, we can write:

$$\frac{T_{e,1}}{T_{e,2}} = \left(\frac{I_{a,1}}{I_{a,2}}\right)^2 \tag{2}$$

As we know, in a series DC machine, the copper loss can be calculated as follows:

$$P_{cu} = \left(R_a + R_s + R_{fc}\right)I_a^2 \Rightarrow \frac{P_{cu,1}}{P_{cu,2}} = \left(\frac{I_{a,1}}{I_{a,2}}\right)^2 \tag{3}$$

Solving (2) and (3):

$$\frac{P_{cu,1}}{P_{cu,2}} = \frac{T_{e,1}}{T_{e,2}} \tag{4}$$

Solving (1) and (4):

$$P_{cu,2} = 2P_{cu,1}$$

Therefore, the copper loss will increase about 100%. Choice (2) is the answer.

18.8 Based on the information given in the problem, we have:

$$N_{s,2} = 0.5N_{s,1} \tag{1}$$

$$R_a = 0 \ \Omega \tag{2}$$

$$\varphi \propto I_a \tag{3}$$

$$T_{load} = \text{Const.} \Rightarrow T_e = \text{Const.} \tag{4}$$

As we know, the electromagnetic torque of an electric machine can be determined by using $T = k_a \varphi I_a$. Moreover, in a series machine with linear magnetization characteristics, we know that $\varphi \propto N_s I_a$ and $I_a = I_s = I_t$. Therefore, we can write:

$$\frac{T_{e,1}}{T_{e,2}} = \frac{N_{s,1} I_{a,1}^2}{N_{s,2} I_{a,2}^2} \tag{5}$$

$$\Rightarrow 1 = \frac{N_{s,1} I_{a,1}^2}{0.5 N_{s,1} I_{a,2}^2} \Rightarrow \frac{I_{a,1}^2}{I_{a,2}^2} = \frac{1}{2} \Rightarrow \frac{I_{a,1}}{I_{a,2}} = \frac{1}{\sqrt{2}} \tag{6}$$

As we know, the induced armature voltage of an electric machine can be determined by using $E_a = k_a \varphi \omega$. In a series machine with linear magnetization characteristics, we know that $\varphi \propto N_s I_a$ and $I_a = I_s = I_t$. Therefore, we can write:

$$\frac{E_{a,1}}{E_{a,2}} = \frac{\varphi_1 \omega_1}{\varphi_2 \omega_2} = \frac{(N_{s,1} I_{a,1}) n_1}{(N_{s,2} I_{a,2}) n_2} \tag{7}$$

$$\Rightarrow \frac{E_{a,1}}{E_{a,2}} = \frac{I_{a,1} n_1}{0.5 I_{a,2} n_2} \tag{8}$$

Since $R_a = 0 \ \Omega$, we have:

$$E_a = V_t \tag{9}$$

Solving (8) and (9):

$$1 = \frac{I_{a,1} n_1}{0.5 \sqrt{2} I_{a,1} n_2} \Rightarrow \frac{n_2}{n_1} = \frac{1}{0.5\sqrt{2}} \Rightarrow \frac{n_2}{n_1} = \sqrt{2}$$

Choice (3) is the answer.

18.9 Based on the information given in the problem, we have:

$$V_t = 200 \ V \tag{1}$$

$$I_{t,1} = 50 \ A \tag{2}$$

$$n_1 = 1000 \ rpm \tag{3}$$

$$R_a = 0.1 \ \Omega \tag{4}$$

$$n_2 = 900 \ rpm \tag{5}$$

$$T_{load} = \text{Const.} \tag{6}$$

$$R_s = 0 \ \Omega \tag{7}$$

$$R_{fc,1} = 0 \ \Omega \tag{8}$$

As we know, the electromagnetic torque of an electric machine can be determined by using $T = k_a \varphi I_a$. In a series machine with linear magnetization characteristics, we know that $\varphi \propto I_a$ and $I_a = I_t$. Therefore, we can write:

$$\frac{T_{e,1}}{T_{e,2}} = \left(\frac{I_{a,1}}{I_{a,2}}\right)^2 \Rightarrow 1 = \left(\frac{I_{a,1}}{I_{a,2}}\right)^2 \Rightarrow I_{a,2} = I_{a,1} \tag{9}$$

Figure 18.5 shows the electrical circuit of a series DC motor. Applying KVL in the loop (the first condition):

$$E_{a,1} = V_t - \left(R_a + R_s + R_{fc,1}\right)I_{a,1} = 200 - (0.1 + 0 + 0) \times 50 = 195\ V \tag{10}$$

As we know, the induced armature voltage of an electric machine can be determined by using $E_a = k_a\varphi\omega_m$. In a series machine with linear magnetization characteristics, we have $\varphi \propto I_a$. Therefore, we can write:

$$\frac{E_{a,1}}{E_{a,2}} = \frac{\varphi_1\omega_1}{\varphi_2\omega_2} = \frac{I_{a,1}n_1}{I_{a,2}n_2} \tag{11}$$

$$\Rightarrow \frac{195}{E_{a,2}} = 1 \times \frac{1000}{900} \Rightarrow E_{a,2} = 175.5\ V \tag{12}$$

Applying KVL in the loop (the second condition):

$$E_{a,2} = V_t - \left(R_a + R_s + R_{fc,2}\right)I_{a,2} \tag{13}$$

$$\Rightarrow 175.5 = 200 - \left(0.1 + 0 + R_{fc,2}\right) \times 50 \Rightarrow R_{fc,2} \simeq 0.4\ \Omega$$

Choice (3) is the answer.

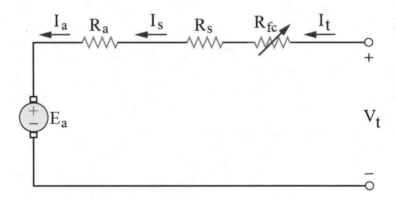

Fig. 18.5 The electrical circuit of solution of problem 18.9

18.10 Based on the information given in the problem, we have:

$$V_t = 400\ V \tag{1}$$

$$I_{t,1} = 50\ A \tag{2}$$

$$n_1 = 1000\ rpm \tag{3}$$

$$R_a = 0.2\ \Omega \tag{4}$$

$$T_{load,2} = 2T_{load,1} \Rightarrow T_{e,2} = 2T_{e,1} \tag{5}$$

$$R_{fc} = 0\ \Omega \tag{6}$$

As we know, the electromagnetic torque of an electric machine can be determined by using $T = k_a\varphi I_a$. In a series machine with linear magnetization characteristics, we know that $\varphi \propto I_a$ and $I_a = I_t$. Therefore, we can write:

$$\frac{T_{e,1}}{T_{e,2}} = \left(\frac{I_{a,1}}{I_{a,2}}\right)^2 \Rightarrow \frac{1}{2} = \left(\frac{50}{I_{a,2}}\right)^2 \Rightarrow I_{a,2} = 50\sqrt{2}\ A \tag{7}$$

Figure 18.6 shows the electrical circuit of a series DC motor. Applying KVL in the loop (the first condition):

$$E_{a,1} = V_t - (R_a + R_s + R_{fc})I_{a,1} = 400 - (0.2 + 0 + 0) \times 50 = 390 \ V \tag{8}$$

Applying KVL in the loop (the second condition):

$$E_{a,2} = V_t - (R_a + R_s + R_{fc})I_{a,2} = 400 - (0.2 + 0 + 0) \times 50\sqrt{2} = 385.85 \ V \tag{9}$$

As we know, the induced armature voltage of an electric machine can be determined by using $E_a = k_a \varphi \omega$. In a series machine with linear magnetization characteristics, we have $\varphi \propto I_a$. Therefore, we can write:

$$\frac{E_{a,1}}{E_{a,2}} = \frac{\varphi_1 \omega_1}{\varphi_2 \omega_2} = \frac{I_{a,1} n_1}{I_{a,2} n_2} \tag{10}$$

$$\Rightarrow \frac{390}{385.85} = \frac{50 \times 1000}{50\sqrt{2} \times n_2} \Rightarrow n_2 \approx 700 \ rpm$$

Choice (4) is the answer.

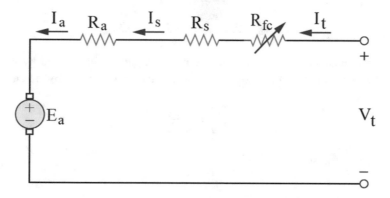

Fig. 18.6 The electrical circuit of solution of problem 18.10

18.11 Based on the information given in the problem, we have:

$$R_f = 0.06 \ \Omega \tag{1}$$

$$I_{t,1} = 20 \ A \tag{2}$$

$$R_{parallel} = 0.12 \ \Omega \tag{3}$$

$$T_{load,2} = 2T_{load,1} \Rightarrow T_{e,2} = 2T_{e,1} \tag{4}$$

As we know, the electromagnetic torque of an electric machine can be determined by using $T = k_a \varphi I_a$. In a series machine with linear magnetization characteristics, we know that $\varphi \propto I_s$. Therefore, we can write:

$$\frac{T_{e,1}}{T_{e,2}} = \frac{I_{s,1} I_{a,1}}{I_{s,2} I_{a,2}} \tag{5}$$

Before adding the new resistor, we have:

$$I_{s,1} = I_{t,1} = I_{a,1} \tag{6}$$

After paralleling the field circuit ($R_f = 0.06\ \Omega$) with the $R_{parallel} = 0.12\ \Omega$, the current of field winding can be determined by using current division rule as follows:

$$I_{s,2} = \frac{0.12}{0.06 + 0.12} I_{a,2} = \frac{2}{3} I_{a,2} \tag{7}$$

Solving (4)–(7):

$$\frac{1}{2} = \frac{I_{a,1}^2}{\frac{2}{3} I_{a,2}^2} \Rightarrow I_{a,2} = \sqrt{3} I_{a,1} \tag{8}$$

Solving (2), (6), and (8):

$$\Rightarrow I_{a,2} = 20\sqrt{3} = 34.6\ A$$

Choice (4) is the answer.

18.12 Based on the information given in the problem, we have:

$$R_{parallel} = 2R_f \tag{1}$$

$$T_{load,2} = 0.5 T_{load,1} \Rightarrow T_{e,2} = 0.5 T_{e,1} \tag{2}$$

As we know, the electromagnetic torque of an electric machine can be determined by using $T = k_a \varphi I_a$. In a series machine with linear magnetization characteristics, we know that $\varphi \propto I_s$. Therefore, we can write:

$$\frac{T_{e,1}}{T_{e,2}} = \frac{I_{s,1} I_{a,1}}{I_{s,2} I_{a,2}} \tag{3}$$

Before adding the new resistor, we have:

$$I_{s,1} = I_{a,1} \tag{4}$$

After paralleling the field circuit resistance (R_f) with the resistor of $2R_f$, the current of field winding can be determined by using current division rule as follows:

$$I_{s,2} = \frac{2R_f}{R_f + 2R_f} I_{a,2} = \frac{2}{3} I_{a,2} \tag{5}$$

Solving (2)–(5):

$$\frac{1}{0.5} = \frac{I_{a,1}^2}{\frac{2}{3} I_{a,2}^2} \Rightarrow I_{a,2} = \frac{\sqrt{3}}{2} I_{a,1} \tag{6}$$

Choice (4) is the answer.

18.13 Based on the information given in the problem, we have:

$$T_{load,2} = 2 T_{load,1} \Rightarrow T_{e,2} = 2 T_{e,1} \tag{1}$$

As we know, the electromagnetic torque of an electric machine can be determined by using $T = k_a \varphi I_a$. In a series machine with linear magnetization characteristics, we know that $\varphi \propto I_s$ and $I_s = I_a$. Therefore, we can write:

$$\frac{T_{e,1}}{T_{e,2}} = \left(\frac{I_{a,1}}{I_{a,2}}\right)^2 \tag{2}$$

Solving (1) and (2):

$$\frac{T_{e,1}}{2T_{e,1}} = \left(\frac{I_{a,1}}{I_{a,2}}\right)^2 \Rightarrow \left(\frac{I_{a,1}}{I_{a,2}}\right)^2 = \frac{1}{2} \tag{3}$$

As we know, the copper loss in a series motor can be calculated as follows:

$$P_{cu} = \left(R_a + R_s + R_{fc}\right)I_a^2 \tag{4}$$

Solving (3) and (4):

$$\frac{P_{cu,1}}{P_{cu,2}} = \left(\frac{I_{a,1}}{I_{a,2}}\right)^2 = \frac{1}{2} \Rightarrow P_{cu,2} = 2P_{cu,1}$$

Choice (1) is the answer.

18.14 The magnetization characteristics of the motor are shown in Fig. 18.7. In addition, based on the information given in the problem, we have:

$$I_{t,FL} = I_{a,FL} = 40 \, A \tag{1}$$

$$T_{e,FL} = 24 \, N.m \tag{2}$$

$$I_{t,start} = I_{a,start} = 100 \, A \tag{3}$$

$$AR = 0 \tag{4}$$

The rated emf can be extracted from the magnetization characteristics of the machine as follows:

$$I_{a,FL} = 40 \, A \Rightarrow \widehat{E}_{a,FL} = 120 \, V \tag{5}$$

The starting emf can be determined by using the magnetization characteristics of the machine as follows:

$$I_{a,start} = 100 \, A \Rightarrow \widehat{E}_{a,start} = 200 \, V \tag{6}$$

As we know, the induced armature voltage of an electric machine can be determined by using $E_a = k_a \varphi \omega$. Since both $\widehat{E}_{a,FL}$ and $\widehat{E}_{a,rated}$ have been extracted from the magnetization characteristics of the machine (no-load condition), the speed of the machine is constant. Therefore:

$$\frac{\widehat{E}_{a,start}}{\widehat{E}_{a,FL}} = \frac{\varphi_{start}}{\varphi_{FL}} \Rightarrow \frac{\varphi_{start}}{\varphi_{FL}} = \frac{200}{120} \tag{7}$$

As we know, the electromagnetic torque of an electric machine can be determined by using $T = k_a \varphi I_a$. Therefore, we can write:

$$\frac{T_{start}}{T_{FL}} = \frac{\varphi_{start}}{\varphi_{FL}} \frac{I_{a,start}}{I_{a,FL}} \tag{8}$$

$$\Rightarrow \frac{T_{start}}{24} = \frac{200}{120} \times \frac{100}{40} \Rightarrow T_{start} = 100 \, N.m$$

Choice (3) is the answer.

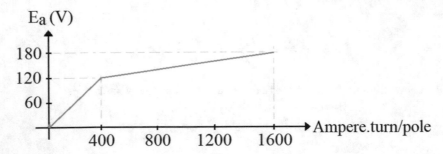

Fig. 18.7 The graph of solution of problem 18.14

18.15 Based on the information given in the problem, we have:

$$n_1 = n_0 \tag{1}$$

$$N_{s,2} = 0.5N_{s,1} \tag{2}$$

$$V_{t,2} = 0.5V_{t,1} \tag{3}$$

$$T_{load,2} = 2T_{load,1} \Rightarrow T_{e,2} = 2T_{e,1} \tag{4}$$

$$R_a = R_s = R_{fc} = 0 \tag{5}$$

As we know, the electromagnetic torque of an electric machine can be determined by using $T = k_a\varphi I_a$. In a series machine with linear magnetization characteristics, we know that $\varphi \propto N_s I_s$ and $I_s = I_a$. Therefore, we can write:

$$\frac{T_{e,1}}{T_{e,2}} = \frac{\varphi_1}{\varphi_2}\frac{I_{a,1}}{I_{a,2}} = \frac{N_{s,1}}{N_{s,2}}\left(\frac{I_{a,1}}{I_{a,2}}\right)^2 \tag{6}$$

$$\Rightarrow \frac{1}{2} = \frac{1}{0.5}\left(\frac{I_{a,1}}{I_{a,2}}\right)^2 \Rightarrow \frac{I_{a,1}}{I_{a,2}} = \frac{1}{2} \tag{7}$$

From (5), we can conclude that:

$$E_a = V_t \tag{8}$$

Solving (3) and (8):

$$\frac{E_{a,1}}{E_{a,2}} = \frac{V_{t,1}}{V_{t,2}} \Rightarrow \frac{E_{a,1}}{E_{a,2}} = \frac{1}{0.5} \tag{9}$$

As we know, the induced armature voltage of an electric machine can be determined by using $E_a = k_a\varphi\omega_m$. In a series machine with linear magnetization characteristics, we have $\varphi \propto N_s I_a$ and $I_s = I_a$. Therefore, we can write:

$$\frac{E_{a,1}}{E_{a,2}} = \frac{\varphi_1\omega_1}{\varphi_2\omega_2} = \frac{N_{s,1}}{N_{s,2}} \times \frac{I_{a,1}}{I_{a,2}} \times \frac{n_1}{n_2} \tag{10}$$

$$\Rightarrow \frac{1}{0.5} = \frac{1}{0.5} \times \frac{1}{2} \times \frac{n_0}{n_2} \Rightarrow n_2 = \frac{1}{2}n_0$$

Choice (2) is the answer.

18.16 Based on the information given in the problem, we have:

$$V_{t,1} = 125 \ V \tag{1}$$

$$n_1 = 1200 \ rpm \tag{2}$$

$$I_{t,1} = 50 \ A \tag{3}$$

$$V_{t,2} = 190 \ V \tag{4}$$

$$T_{load,2} = 2\sqrt{2}T_{load,1} \Rightarrow T_{e,2} = 2\sqrt{2}T_{e,1} \tag{5}$$

$$AR = 0 \tag{6}$$

$$R_a = 0.1 \ \Omega \tag{7}$$

$$\varphi \propto \sqrt{I_a} \tag{8}$$

$$R_s = R_{fc} = 0 \ \Omega \tag{9}$$

As we know, the electromagnetic torque of an electric machine can be determined by using $T = k_a\varphi I_a$. In a series machine with linear magnetization characteristics, we know that $\varphi \propto I_s$ and $I_s = I_a = I_t$. Therefore, we can write:

$$\frac{T_{e,1}}{T_{e,2}} = \frac{\varphi_1}{\varphi_2}\frac{I_{a,1}}{I_{a,2}} \tag{10}$$

$$\Rightarrow \frac{T_{e,1}}{T_{e,2}} = \frac{\sqrt{I_{a,1}}}{\sqrt{I_{a,2}}}\frac{I_{a,1}}{I_{a,2}} \Rightarrow \frac{1}{2\sqrt{2}} = \left(\frac{50}{I_{a,2}}\right)^{\frac{3}{2}} \Rightarrow I_{a,2} = 100 \ A \tag{11}$$

Figure 18.8 shows the electrical circuit of a series DC motor. Applying KVL in the loop (the first condition):

$$E_{a,1} = V_t - \left(R_a + R_s + R_{fc}\right)I_{a,1} = 125 - (0.1 + 0 + 0) \times 50 = 120 \ V \tag{12}$$

Applying KVL in the loop (the second condition):

$$E_{a,2} = V_t - \left(R_a + R_s + R_{fc}\right)I_{a,2} = 190 - (0.1 + 0 + 0) \times 100 = 180 \ V \tag{13}$$

As we know, the induced armature voltage of an electric machine can be determined by using $E_a = k_a\varphi\omega$. Therefore, we can write:

$$\frac{E_{a,1}}{E_{a,2}} = \frac{\varphi_1\omega_1}{\varphi_2\omega_2} = \sqrt{\frac{I_{a,1}}{I_{a,2}}}\frac{n_1}{n_2} \tag{14}$$

$$\Rightarrow \frac{120}{180} = \sqrt{\frac{50}{100}} \times \frac{1200}{n_2} \Rightarrow n_2 = 900\sqrt{2} \ rpm$$

Choice (2) is the answer.

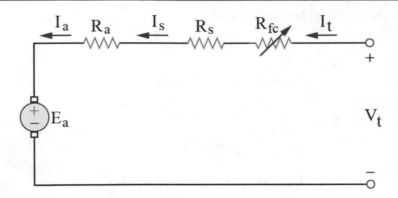

Fig. 18.8 The electrical circuit of solution of problem 18.16

18.17 The magnetization characteristics of a series DC motor are shown in Fig. 18.9.a. Moreover, based on the information given in the problem, we have:

$$N_s = 25 \text{ turns/pole} \tag{1}$$

$$R_a + R_s + R_{fc} = 2 \ \Omega \tag{2}$$

$$V_{t,start} = 128 \ V \tag{3}$$

$$I_{t,rated} = 12 \ A \tag{4}$$

Figure 18.9.b shows the electrical circuit of a series DC motor. The starting current of motor can be calculated as follows:

$$I_{t,start} = \frac{V_{t,start}}{R_a + R_s + R_{fc}} = \frac{128}{2} = 64 \ A \tag{5}$$

The starting ampere-turn per pole (AT/pole) of machine can be calculated as follows:

$$\Rightarrow \text{Ampere} - \text{turn per pole}|_{start} = N_s I_{t,start} = 25 \times 64 = 1600 \ \frac{AT}{P} \tag{6}$$

The starting emf can be determined by using the magnetization characteristics of machine as follows:

$$\text{Ampere turn per pole}|_{start} = 1600 \ \frac{AT}{P} \Rightarrow \widehat{E}_{a,start} = 180 \ V \tag{7}$$

The rated ampere-turn per pole (AT/pole) of machine can be calculated as follows:

$$\text{Ampere turn per pole}|_{rated} = N_s I_{t,rated} = 25 \times 12 = 300 \ \frac{AT}{P} \tag{8}$$

The rated emf can be extracted from the magnetization characteristics of the machine as follows:

$$\text{Ampere} - \text{turn per pole}|_{rated} = 300 \ \frac{AT}{P} \Rightarrow \widehat{E}_{a,rated} = 90 \ V \tag{9}$$

As we know, the induced armature voltage of an electric machine can be determined by using $E_a = k_a \varphi \omega$. Since both $\widehat{E}_{a,start}$ and $\widehat{E}_{a,rated}$ have been extracted from the magnetization characteristics of machine (no-load condition), the speed is constant. Therefore:

$$\frac{\widehat{E}_{a,start}}{\widehat{E}_{a,rated}} = \frac{\varphi_{start}}{\varphi_{rated}} = \frac{180}{90} = 2 \tag{10}$$

As we know, the electromagnetic torque of an electric machine can be determined by using $T = k_a \varphi I_a$. Therefore, we can write:

$$\frac{T_{start}}{T_{rated}} = \frac{\varphi_{start} I_{start}}{\varphi_{rated} I_{rated}} \tag{11}$$

$$\Rightarrow \frac{T_{start}}{T_{rated}} = 2 \times \frac{64}{12} \Rightarrow \frac{T_{start}}{T_{rated}} = 10.67$$

Choice (4) is the answer.

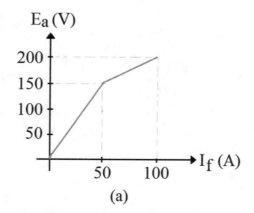

(a)

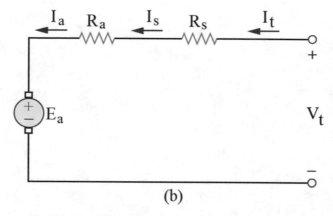

(b)

Fig. 18.9 The graph and electrical circuit of solution of problem 18.17

Problems: Compound DC Electric Motor

Abstract

In this chapter, the problems concerned with the compound DC electric motors are solved. In this chapter, the problems are categorized in different levels based on their difficulty levels (easy, normal, and hard) and calculation amounts (small, normal, and large). Additionally, the problems are ordered from the easiest problem with the smallest computations to the most difficult problems with the largest calculations.

19.1 A 250 V, 86.25 A, 720 rpm, long-shunt cumulative compound DC motor has the current and speed of 5.75 A and 1150 rpm when it is not connected to any load. The shunt field winding resistance, armature winding resistance, and series field winding resistance are 200 Ω, 0.15 Ω, and 0.05 Ω, respectively. Calculate the ratio of magnetic flux in the full-load condition to the no-load one.

Difficulty level ○ Easy ● Normal ○ Hard
Calculation amount ○ Small ● Normal ○ Large
1) 0.668
2) 1.292
3) 1.321
4) 1.494

19.2 A 5 kW, 200 V, 1000 rpm, long-shunt flat compound DC generator has the armature winding, series field winding, and shunt field winding resistances of 0.05 Ω, 0.05 Ω, and 200 Ω, respectively. If the machine is operated as a motor, calculate its rated speed considering the fact that the magnetic flux is the same in both generating and motor actions.

Difficulty level ○ Easy ● Normal ○ Hard
Calculation amount ○ Small ● Normal ○ Large
1) 1000 *rpm*
2) 925 *rpm*
3) 975 *rpm*
4) 950 *rpm*

19.3 A long-shunt compound DC motor provides the torque of 10 N.m with the armature current of 20 A and the shunt field current of 4 A. Determine the ratio of the number of turns of series field winding to the ones of shunt field winding if this motor, in another condition, provides the torque of 15 N.m with the armature current of 30 A and the shunt field current of 3 A. Assume a linear magnetic circuit for the motor and ignore the armature reaction.

Difficulty level ○ Easy ○ Normal ● Hard
Calculation amount ○ Small ● Normal ○ Large
1) 0.01
2) 0.15
3) 0.1
4) 0.2

© The Author(s), under exclusive license to Springer Nature Switzerland AG 2022
M. Rahmani-Andebili, *DC Electric Machines, Electromechanical Energy Conversion Principles, and Magnetic Circuit Analysis*,
https://doi.org/10.1007/978-3-031-08863-6_19

19.4 In a compound DC motor, the torque, the series field winding current, and the shunt field winding current are 1.25 p.u., 0.8 p.u., and 1.4 p.u., respectively, and the armature current is rated. Determine the ratio of the number of turns of series field winding to the ones of shunt field winding.

Difficulty level   ○ Easy   ○ Normal   ● Hard
Calculation amount   ○ Small   ● Normal   ○ Large

1) $\frac{1}{2}$

2) $\frac{1}{4}$

3) $\frac{1}{5}$

4) $\frac{1}{3}$

19.5 A shunt DC motor has a linear magnetic circuit and its power losses are negligible. When the motor is supplied by 200 V, its armature current and speed are 50 A and n_0. Now, the motor with the same load torque is changed to a cumulative compound DC motor by adding a series winding to it with the number of turns of N_s per pole. This modified motor is supplied by 240 V and its speed becomes $0.8n_0$. Determine the value of N_s if the number of turns of the shunt field winding is 1000 per pole and its resistance is 50 Ω.

Difficulty level   ○ Easy   ○ Normal   ● Hard
Calculation amount   ○ Small   ○ Normal   ● Large

1) 16

2) 24

3) 36

4) 60

Solutions of Problems: Compound DC Electric Motor

20

Abstract

In this chapter, the problems of the 19th chapter are fully solved, in detail, step-by-step, and with different methods.

20.1 Based on the information given in the problem, we have:

$$V_t = 250 \ V \tag{1}$$

$$I_{t,FL} = 86.25 \ A \tag{2}$$

$$n_{FL} = 720 \ rpm \tag{3}$$

$$I_{t,NL} = 5.75 \ A \tag{4}$$

$$n_{NL} = 1150 \ rpm \tag{5}$$

$$R_f = 200 \ \Omega \tag{6}$$

$$R_a = 0.15 \ \Omega \tag{7}$$

$$R_s = 0.05 \ \Omega \tag{8}$$

Figure 20.1 shows the electrical circuit of a long-shunt compound DC motor. Applying Ohm's law for the branch of field winding:

$$I_f = \frac{V_t}{R_f} = \frac{V_t}{R_{fc} + R_{fw}} = \frac{250}{200} = 1.25 \ A \tag{9}$$

Applying KCL at the top node of the circuit in the full-load condition:

$$I_{a,FL} = I_{t,FL} - I_f = 86.25 - 1.25 = 85 \ A \tag{10}$$

Applying KCL at the top node of the circuit in the no-load condition:

$$I_{a,NL} = I_{t,NL} - I_f = 5.75 - 1.25 = 4.5 \ A \tag{11}$$

Applying KVL in the right-hand side mesh of the circuit in the full-load condition:

$$E_{a,FL} = V_t - (R_a + R_s)I_{a,FL} = 250 - (0.15 + 0.05) \times 85 = 233 \ V \tag{12}$$

© The Author(s), under exclusive license to Springer Nature Switzerland AG 2022
M. Rahmani-Andebili, *DC Electric Machines, Electromechanical Energy Conversion Principles, and Magnetic Circuit Analysis*,
https://doi.org/10.1007/978-3-031-08863-6_20

Applying KVL in the right-hand side mesh of the circuit in the no-load condition:

$$E_{a,NL} = V_t - (R_a + R_s)I_{a,NL} = 250 - (0.15 + 0.05) \times 4.5 = 249.1 \ V \tag{13}$$

As we know, the induced armature voltage of an electric machine can be determined by using $E_a = k_a \varphi \omega$. Therefore, we can write:

$$\frac{E_{a,FL}}{E_{a,NL}} = \frac{\varphi_{FL}\omega_{FL}}{\varphi_{NL}\omega_{NL}} = \frac{\varphi_{FL}}{\varphi_{NL}} \times \frac{n_{FL}}{n_{NL}} \tag{14}$$

$$\Rightarrow \frac{233}{249.1} = \frac{\varphi_{FL}}{\varphi_{NL}} \times \frac{720}{1150} \Rightarrow \frac{\varphi_{FL}}{\varphi_{NL}} = 1.494$$

Choice (4) is the answer.

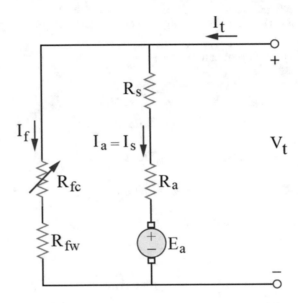

Fig. 20.1 The electrical circuit of solution of problem 20.1

20.2 Based on the information given in the problem, we have:

$$P = 5 \ kW \tag{1}$$

$$V_t = 200 \ V \tag{2}$$

$$n_g = 1000 \ rpm \tag{3}$$

$$R_a = 0.05 \ \Omega \tag{4}$$

$$R_s = 0.05 \ \Omega \tag{5}$$

$$R_f = 200 \ \Omega \tag{6}$$

$$\varphi_m = \varphi_g \tag{7}$$

The rated terminal current of machine can be calculated as follows:

$$I_{t,g} = I_{t,m} = \frac{P}{V_t} = \frac{5000}{200} = 25 \ A \tag{8}$$

Figure 20.2 shows the electrical circuit of a long-shunt compound DC motor. Applying Ohm's law for the branch of field winding:

$$I_{f,g} = I_{f,m} = \frac{V_t}{R_f} = \frac{V_t}{R_{fc} + R_{fw}} = \frac{200}{200} = 1 \, A \tag{9}$$

Applying KCL at the top node of the circuit in the generating condition:

$$\Rightarrow I_{a,g} = I_{t,g} + I_{f,g} = 25 + 1 = 26 \, A \tag{10}$$

Applying KVL in the right-hand side mesh of the circuit in the generating condition:

$$E_{a,g} = V_t + (R_a + R_s)I_{a,g} = 200 + (0.05 + 0.05) \times 26 = 202.6 \, V \tag{11}$$

Applying KCL at the top node of the circuit in the motoring condition:

$$I_{a,m} = I_{t,m} - I_{f,m} = 25 - 1 = 24 \, A \tag{12}$$

Applying KVL in the right-hand side mesh of the circuit in the motoring condition:

$$E_{a,m} = V_t - (R_a + R_s)I_{a,m} = 200 - (0.05 + 0.05) \times 24 = 197.6 \, V \tag{13}$$

As we know, the induced armature voltage of an electric machine can be determined by using $E_a = k_a \varphi \omega$. Therefore, we can write:

$$\frac{E_{a,g}}{E_{a,m}} = \frac{\omega_g \varphi_g}{\omega_m \varphi_m} = \frac{n_g \varphi_g}{n_m \varphi_m} \tag{14}$$

Solving (7) and (14):

$$\frac{E_{a,g}}{E_{a,m}} = \frac{n_g}{n_m} \tag{15}$$

$$\Rightarrow \frac{202.6}{197.6} = \frac{1000}{n_m} \Rightarrow n_m = 975 \, rpm$$

Choice (3) is the answer.

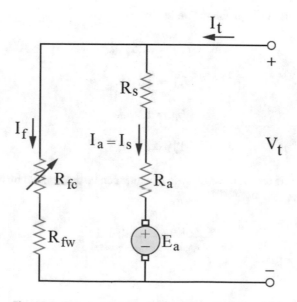

Fig. 20.2 The electrical circuit of solution of problem 20.2

20.3 Based on the information given in the problem, we have:

$$T_{e,1} = 10 \, N.m \tag{1}$$

$$I_{a,1} = 20 \, A \tag{2}$$

$$I_{f,1} = 4 \, A \tag{3}$$

$$T_{e,2} = 15 \, N.m \tag{4}$$

$$I_{a,2} = 30 \, A \tag{5}$$

$$I_{f,2} = 3 \, A \tag{6}$$

$$AR = 0 \tag{7}$$

As we know, the electromagnetic torque of an electric machine can be determined by using $T = k_a \varphi I_a$. Moreover, the total magnetic flux is sum of the magnetic flux of the series and shunt field windings. Therefore, we can write:

$$\frac{T_{e,1}}{T_{e,2}} = \frac{\varphi_1 I_{a,1}}{\varphi_2 I_{a,2}} \Rightarrow \frac{T_{e,1}}{T_{e,2}} = \frac{\varphi_{f,1} + \varphi_{s,1}}{\varphi_{f,2} + \varphi_{s,2}} \times \frac{I_{a,1}}{I_{a,2}} \tag{8}$$

In addition, since the magnetic circuit of motor is linear, we can use the ampere-turn of the windings instead of their magnetic fluxes as follows:

$$\frac{T_{e,1}}{T_{e,2}} = \frac{\left(N_{sh} I_{f,1} + N_s I_{a,1}\right) I_{a,1}}{\left(N_{sh} I_{f,2} + N_s I_{a,2}\right) I_{a,2}} \tag{9}$$

$$\Rightarrow \frac{10}{15} = \frac{(4N_{sh} + 20N_s)20}{(3N_{sh} + 30N_s)30} \Rightarrow 4N_{sh} + 20N_s = 3N_{sh} + 30N_s \Rightarrow N_{sh} = 10N_s$$

$$\Rightarrow \frac{N_s}{N_{sh}} = 0.1$$

Choice (3) is the answer.

20.4 Based on the information given in the problem, we have:

$$T_{e,1} = 1.25 \, p.u. \tag{1}$$

$$I_{s,1} = 0.8 \, p.u. \tag{2}$$

$$I_{f,1} = 1.4 \, p.u. \tag{3}$$

$$I_{a,1} = 1 \, p.u. \tag{4}$$

As we know, the electromagnetic torque of an electric machine can be determined by using $T = k_a \varphi I_a$. Therefore, we can write:

$$\frac{T_{e,1}}{T_{e,rated}} = \frac{\varphi_1 I_{a,1}}{\varphi_{rataed} I_{a,rated}} \tag{5}$$

Since the magnetic circuit of machine is linear, we can use the ampere-turn of the windings instead of their magnetic fluxes as follows:

$$\frac{\varphi_1}{\varphi_{rated}} = \frac{N_{sh}I_{f,1} + N_s I_{a,1}}{N_{sh}I_{f,rated} + N_s I_{a,rated}} \tag{6}$$

Solving (5) and (6):

$$\frac{T_{e,1}}{T_{e,rated}} = \frac{N_{sh}I_{f,1} + N_s I_{a,1}}{N_{sh}I_{f,rated} + N_s I_{a,rated}} \times \frac{I_{a,1}}{I_{a,rated}} \tag{7}$$

$$\Rightarrow \frac{1.25}{1} = \frac{(N_{sh} \times 1.4) + (N_s \times 0.8)}{(N_{sh} \times 1) + (N_s \times 1)} \times \frac{1}{1} \Rightarrow 1.25N_{sh} + 1.25N_s = 1.4N_{sh} + 0.8N_s$$

$$\Rightarrow 0.45N_s = 0.15N_{sh} \Rightarrow \frac{N_s}{N_{sh}} = \frac{1}{3}$$

Choice (4) is the answer.

20.5 Based on the information given in the problem, we have:

$$R_a = R_s = 0\ \Omega \tag{1}$$

$$V_{t,1} = 200\ V \tag{2}$$

$$I_{a,1} = 50\ A \tag{3}$$

$$n_1 = n_0 \tag{4}$$

$$T_{load,1} = T_{load,2} \Rightarrow T_{e,1} = T_{e,2} \tag{5}$$

$$V_{t,2} = 240\ V \tag{6}$$

$$n_1 = 0.8n_0 \tag{7}$$

$$N_{sh} = 1000\ \text{turns/pole} \tag{8}$$

$$R_f = 50\ \Omega \tag{9}$$

Figure 20.3.a and Fig. 20.3.b show the electrical circuits of a shunt DC motor and a long-shunt compound DC motor. Applying Ohm's law for the branch of field winding:

$$I_{f,1} = \frac{V_{t,1}}{R_f} = \frac{V_{t,1}}{R_{fc} + R_{fw}} = \frac{200}{50} = 4\ A \tag{10}$$

$$I_{f,2} = \frac{V_{t,2}}{R_f} = \frac{V_{t,2}}{R_{fc} + R_{fw}} = \frac{240}{50} = 4.8\ A \tag{11}$$

As we know, the electromagnetic torque of an electric machine can be determined by using $T = k_a \varphi I_a$. Therefore, we can write:

$$\frac{T_{e,1}}{T_{e,2}} = \frac{\varphi_1 I_{a,1}}{\varphi_2 I_{a,2}} \Rightarrow 1 = \frac{\varphi_1 I_{a,1}}{\varphi_2 I_{a,2}} \Rightarrow \frac{\varphi_1}{\varphi_2} = \frac{I_{a,2}}{I_{a,1}} \tag{12}$$

As we know, the induced armature voltage of an electric machine can be determined by using $E_a = k_a \varphi \omega$. Therefore, we can write:

$$\frac{E_{a,1}}{E_{a,2}} = \frac{\varphi_1 \omega_1}{\varphi_2 \omega_2} = \frac{\varphi_1 n_1}{\varphi_2 n_2} \tag{13}$$

Solving (12) and (13):

$$\frac{E_{a,1}}{E_{a,2}} = \frac{I_{a,2} n_1}{I_{a,1} n_2} \tag{14}$$

Since the power losses are negligible ($R_a = R_s = 0 \ \Omega$), we have:

$$E_a = V_t \tag{15}$$

Solving (14) and (15):

$$\frac{V_{t,1}}{V_{t,2}} = \frac{I_{a,2} n_1}{I_{a,1} n_2} \tag{16}$$

$$\Rightarrow \frac{200}{240} = \frac{I_{a,2}}{50} \times \frac{n_0}{0.8 n_0} \Rightarrow I_{a,2} = \frac{100}{3} \ A \tag{17}$$

Since the magnetic circuit of the machine is linear, we can use the ampere-turn of the windings instead of their magnetic fluxes as follows:

$$\frac{\varphi_1}{\varphi_2} = \frac{N_{sh} I_{f,1}}{N_{sh} I_{f,2} + N_s I_{a,2}} \tag{18}$$

Solving (12) and (18):

$$\frac{I_{a,2}}{I_{a,1}} = \frac{N_{sh} I_{f,1}}{N_{sh} I_{f,2} + N_s I_{a,2}} \tag{19}$$

$$\Rightarrow \frac{\frac{100}{3}}{50} = \frac{1000 \times 4}{(1000 \times 4.8) + \left(N_s \times \frac{100}{3}\right)} \Rightarrow N_s \simeq 36 \text{ turns}$$

Choice (3) is the answer.

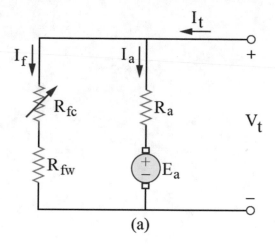

(a)

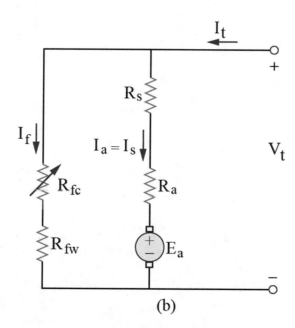

(b)

Fig. 20.3 The electrical circuits of solution of problem 20.5

Problems: Power Loss and Efficiency of DC Electric Machines

21

Abstract

In this chapter, the problems concerned with the power loss and energy efficiency of DC electric generators and motors are solved. In this chapter, the problems are categorized in different levels based on their difficulty levels (easy, normal, and hard) and calculation amounts (small, normal, and large). Additionally, the problems are ordered from the easiest problem with the smallest computations to the most difficult problems with the largest calculations.

21.1 In a 100 V permanent magnet DC generator, the armature winding resistance and the rotational power loss are 0.1 Ω and 90 W, respectively. Calculate the power in which the efficiency is maximum.

Difficulty level ○ Easy ● Normal ○ Hard
Calculation amount ● Small ○ Normal ○ Large
1) 2 kW
2) 4 kW
3) 3 kW
4) 3.5 kW

21.2 A 5 hp, 220 V, shunt DC motor has the efficiency of 80% at its rated speed. If this machine is operated as a generator with the same voltage, line current, and speed, calculate its efficiency.

Difficulty level ○ Easy ● Normal ○ Hard
Calculation amount ○ Small ● Normal ○ Large
1) 76.7%
2) 80%
3) 83.3%
4) 86.6%

21.3 In a 100 V permanent magnet DC generator, the armature winding resistance and the rotational power loss are 0.1 Ω and 40 W, respectively. Calculate the maximum efficiency of generator.

Difficulty level ○ Easy ● Normal ○ Hard
Calculation amount ○ Small ● Normal ○ Large
1) 94%
2) 90%
3) 96%
4) 80%

21.4 A 200 V, 100 A, shunt DC generator is rotating by an internal combustion engine with the output power of 32 hp, as the prime mover. The armature winding resistance and field winding resistance of generator are 0.1 Ω and 50 Ω, respectively. Calculate the sum of the core and rotational losses of the machine.

Difficulty level ○ Easy ● Normal ○ Hard
Calculation amount ○ Small ● Normal ○ Large

© The Author(s), under exclusive license to Springer Nature Switzerland AG 2022
M. Rahmani-Andebili, *DC Electric Machines, Electromechanical Energy Conversion Principles, and Magnetic Circuit Analysis*,
https://doi.org/10.1007/978-3-031-08863-6_21

1) 995.2 *W*

2) 1200.4 *W*

3) 1990.4 *W*

4) 1881.4 *W*

21.5 In a 5 kW, 200 V shunt DC generator, the field winding resistance is 100 Ω. If the armature reaction decreases by 5% of the armature voltage, calculate the rated copper loss of machine.

Difficulty level   ○ Easy   ● Normal   ○ Hard
Calculation amount   ○ Small   ● Normal   ○ Large

1) 270 *W*

2) 770 *W*

3) 670 *W*

4) 570 *W*

21.6 In a series DC motor, the rotational power loss is about 1% of the input power. Calculate the efficiency of this motor for its maximum output power.

Difficulty level   ○ Easy   ○ Normal   ● Hard
Calculation amount   ○ Small   ● Normal   ○ Large

1) 49%

2) 69%

3) 79%

4) 89%

21.7 A permanent magnet DC motor has the rated quantities below.

$$V_t = 300 \ V, \ I_t = 50 \ A, \ P = 12 \ kW, \ n = 2400 \ rpm$$

Calculate its full-load efficiency and no-load speed if its rotational power loss is negligible.

Difficulty level   ○ Easy   ○ Normal   ● Hard
Calculation amount   ○ Small   ○ Normal   ● Large

1) 80%, 2600 *rpm*

2) 85%, 2800 *rpm*

3) 90%, 3000 *rpm*

4) 80%, 3000 *rpm*

21.8 A series DC motor has the rated speed of 1000 rpm. The ohmic power loss of machine is about 3% and the rotational power loss is constant and about 2%. Calculate the speed of motor when its efficiency is maximum. Herein, ignore the saturation effect.

Difficulty level   ○ Easy   ○ Normal   ● Hard
Calculation amount   ○ Small   ○ Normal   ● Large

1) 1000 *rpm*

2) 1176 *rpm*

3) 1205 *rpm*

4) 1232 *rpm*

Solutions of Problems: Power Loss and Efficiency of DC Electric Machines
22

Abstract

In this chapter, the problems of the 21st chapter are fully solved, in detail, step-by-step, and with different methods.

22.1 Based on the information given in the problem, we have:

$$V_t = 100 \, V \tag{1}$$

$$R_a = 0.1 \, \Omega \tag{2}$$

$$P_{rot} = 90 \, W \tag{3}$$

An electric machine has the maximum efficiency if its variable power loss is equal to its constant power loss. Therefore, for a permanent magnet DC machine, we have:

$$P_{cu}|_{\eta_{max}} = P_{rot} \Rightarrow P_{rot} = R_a I_a^2 \tag{4}$$

$$\Rightarrow 90 = 0.1 I_a^2 \Rightarrow I_a = 30 \, A \tag{6}$$

The output power of the generator can be calculated as follows:

$$P_{out} = V_t I_t = V_t I_a \tag{7}$$

$$P_{out} = 100 \times 30 \Rightarrow P_{out} = 3 \, kW$$

Choice (3) is the answer.

22.2 Based on the information given in the problem, we have:

$$P_{m,out} = 5 \, hp = 5 \times 746 = 3730 \, W \tag{1}$$

$$\eta_m = 80\% \tag{2}$$

$$V_{t,m} = V_{t,g} = 220 \, V \tag{3}$$

$$I_{t,m} = I_{t,g} \tag{4}$$

$$n_m = n_g \tag{5}$$

© The Author(s), under exclusive license to Springer Nature Switzerland AG 2022
M. Rahmani-Andebili, *DC Electric Machines, Electromechanical Energy Conversion Principles, and Magnetic Circuit Analysis*,
https://doi.org/10.1007/978-3-031-08863-6_22

The input power of motor can be calculated using the efficiency formula as follows:

$$\eta_m = \frac{P_{m,out}}{P_{m,in}} \times 100 \Rightarrow P_{m,in} = \frac{3730}{80} \times 100 = 4662.5 \ W \tag{6}$$

The power loss of motor can be calculated as follows:

$$P_{m,loss} = P_{m,in} - P_{m,out} = 4662.5 - 3730 = 932.5 \ W \tag{7}$$

As we know, the relations below exist for the machine, since the speed, terminal voltage, terminal current, armature current, and field current in generating operation are the same as the ones in motoring operation:

$$P_{g,out} = P_{m,in} = 4662.5 \tag{8}$$

$$P_{g,loss} = P_{m,loss} = 932.5 \tag{9}$$

Thus, the efficiency of generating mode can be calculated as follows:

$$\eta_g = \frac{P_{g,out}}{P_{g,out} + P_{g,loss}} \times 100 \tag{10}$$

$$\Rightarrow \eta_g = \frac{4662.5}{4662.5 + 932.5} \times 100 \Rightarrow \eta_g = 83.3\%$$

Choice (3) is the answer.

22.3 Based on the information given in the problem, we have:

$$V_t = 100 \ V \tag{1}$$

$$R_a = 0.1 \ \Omega \tag{2}$$

$$P_{rot} = 40 \ W \tag{3}$$

As we know, an electric machine has the maximum efficiency if its variable power loss is equal to its constant power loss. Therefore, for a permanent magnet DC machine, we have:

$$P_{cu}|_{\eta_{max}} = P_{rot} \tag{4}$$

$$\Rightarrow P_{rot} = R_a I_a^2 \tag{5}$$

$$\Rightarrow 40 \ W = 0.1 I_a^2 \Rightarrow I_a = 20 \ A \tag{6}$$

The output power of generator can be calculated as follows:

$$P_{out} = V_t I_t = V_t I_a = 100 \times 20 = 2000 \ W \tag{7}$$

The input power of generator can be calculated as follows:

$$P_{in} = P_{out} + P_{cu}|_{\eta_{max}} + P_{rot} \tag{8}$$

Solving (4) and (8):

$$P_{in} = P_{out} + 2P_{rot} = 2000 + (2 \times 40) = 2080 \ W \tag{9}$$

The efficiency of generator can be calculated as follows:

$$\eta_{max} = \frac{P_{out}}{P_{in}} \times 100 \tag{10}$$

$$\Rightarrow \eta_{max} = \frac{2000}{2080} \times 100 \Rightarrow \eta_{max} = 96\%$$

Choice (3) is the answer.

22.4 Based on the information given in the problem, we have:

$$V_t = 200 \ V \tag{1}$$

$$I_t = 100 \ A \tag{2}$$

$$P_{in} = 32 \ hp \tag{3}$$

$$R_a = 0.1 \ \Omega \tag{4}$$

$$R_f = 50 \ \Omega \tag{5}$$

Figure 22.1 shows the electrical circuit of a shunt DC generator. Applying Ohm's law for the branch of field winding:

$$I_f = \frac{V_t}{R_{fc} + R_{fw}} = \frac{V_t}{R_f} = \frac{200}{50} = 4 \ A \tag{6}$$

Applying KCL at the top node of the circuit:

$$I_a = I_t + I_f = 100 + 4 = 104 \ A \tag{7}$$

The total power loss of generator can be calculated as follows:

$$P_{loss} = P_{in} - P_{out} = P_{in} - V_t I_t = (32 \times 746) - (200 \times 100) = 3872 \ W \tag{8}$$

On the other hand, the total power loss of machine can be calculated by using another relation as follows:

$$P_{loss} = R_a I_a^2 + (R_{fc} + R_{fw}) I_f^2 + P_{rot} + P_c = R_a I_a^2 + R_f I_f^2 + P_{rot} + P_c \tag{9}$$

Solving (8) and (9) and considering the value of parameters:

$$3872 = (0.1 \times 104^2) + (50 \times 4^2) + P_{rot} + P_c$$

$$\rightarrow P_{rot} + P_c = 1990.4 \ W$$

Choice (3) is the answer.

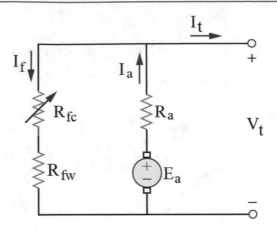

Fig. 22.1 The electrical circuit of solution of problem 22.4

22.5 Based on the information given in the problem, the armature reaction decreases by 5% of the armature voltage. Moreover, we have:

$$P = 5\ kW \tag{1}$$

$$V_t = 200\ V \tag{2}$$

$$R_f = 100\ \Omega \tag{3}$$

$$R_a = 0\ \Omega \tag{4}$$

Figure 22.2 shows the electrical circuit of a shunt DC generator. Applying Ohm's law for the branch of field winding:

$$I_f = \frac{V_t}{R_{fc} + R_{fw}} = \frac{V_t}{R_f} = \frac{200}{100} = 2\ A \tag{5}$$

The terminal current of machine can be calculated as follows:

$$I_t = \frac{P_{out}}{V_t} = \frac{5000}{200} = 25\ A \tag{6}$$

Applying KCL at the top node of the circuit:

$$I_a = I_L + I_f = 25 + 2 = 27\ A \tag{7}$$

By applying KVL in the right-side mesh of the circuit, we have the following relation. Herein, ε is the voltage drop due to the armature reaction:

$$E_a = V_t + R_a I_a + \varepsilon = V_t + 0 + 0.05 V_t = 1.05 V_t = 1.05 \times 200 = 210\ V \tag{8}$$

The electromagnetic power developed can be calculated as follows:

$$P_e = E_a I_a = 210 \times 27 = 5670\ W \tag{9}$$

$$P_{cu} = P_e - P_{out} = 5670 - 5000 \Rightarrow P_{cu} = 670\ W$$

Choice (3) is the answer.

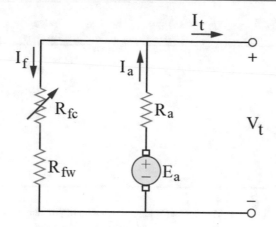

Fig. 22.2 The electrical circuit of solution of problem 22.5

22.6 Based on the information given in the problem, we have:

$$P_{rot} = 0.01P_{in} \tag{1}$$

Figure 22.3 shows the electrical circuit of a series DC motor. The input and output powers of a series DC motor can be determined by the relation below:

$$P_{in} = V_t I_t = V_t I_a \tag{2}$$

$$P_{out} = P_{in} - P_{cu} - P_{rot} = V_t I_a - (R_a + R_s)I_a^2 - P_{rot} \tag{3}$$

To have maximum output power, we need to calculate the first derivate of the output power with respect to its variable, that is, the armature current:

$$\frac{dP_{out}}{dI_a} = 0 \Rightarrow \frac{d}{dI_a}\left(V_t I_a - (R_a + R_s)I_a^2 - P_{rot}\right) = 0 \tag{4}$$

$$\Rightarrow V_t - 2(R_a + R_s)I_a = 0 \Rightarrow I_a = \frac{V_t}{2(R_a + R_s)} \tag{5}$$

Now, the efficiency of motor for its maximum output power can be calculated as follows:

$$\eta = \frac{P_{out}}{P_{in}} \times 100 = \frac{V_t I_a - (R_a + R_s)I_a^2 - P_{rot}}{V_t I_a} \times 100 = \left(1 - \frac{(R_a + R_s)I_a}{V_t} - \frac{P_{rot}}{V_t I_a}\right) \times 100 \tag{6}$$

Solving (2), (5), and (6):

$$\eta = \left(1 - \frac{R_a + R_s}{V_t} \times \left(\frac{V_t}{2(R_a + R_s)}\right) - \frac{P_{rot}}{P_{in}}\right) \times 100 = \left(1 - \frac{1}{2} - \frac{P_{rot}}{P_{in}}\right) \times 100 \tag{7}$$

Solving (1) and (7):

$$\Rightarrow \eta = \left(1 - \frac{1}{2} - 0.01\right) \times 100 \Rightarrow \eta = 49\%$$

Choice (1) is the answer.

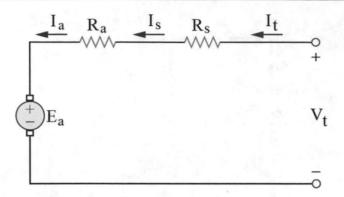

Fig. 22.3 The electrical circuit of solution of problem 22.6

22.7 Based on the information given in the problem, we have:

$$V_t = 300 \ V \tag{1}$$

$$I_{t,FL} = I_{a,FL} = 50 \ A \tag{2}$$

$$P_{out,FL} = 12 \ kW \tag{3}$$

$$n_{FL} = 2400 \ rpm \tag{4}$$

$$P_{rot} = 0 \ W \tag{5}$$

The input power of motor can be calculated as follows:

$$P_{in,FL} = V_t I_{a,FL} = 300 \times 50 = 15000 \ W \tag{6}$$

The efficiency of motor can be calculated as follows:

$$\Rightarrow \eta_{FL} = \frac{P_{out,FL}}{P_{in,FL}} \times 100 \tag{7}$$

$$\Rightarrow \eta_{FL} = \frac{12000}{15000} \times 100 \Rightarrow \eta_{FL} = 80\% \tag{8}$$

By using the output power formula of motor, we have:

$$P_{out,FL} = P_{e,FL} - P_{rot} = E_{a,FL} I_{a,FL} - P_{rot} \tag{9}$$

$$\Rightarrow 12000 \ W = E_{a,FL} I_{a,FL} - 0 \Rightarrow E_{a,FL} = \frac{12000}{50} = 240 \ V \tag{10}$$

As we know, at no-load condition, we have:

$$E_{a,NL} = V_t = 300 \ V \tag{11}$$

As we know, the induced armature voltage of an electric machine can be determined by using $E_a = k_a \varphi \omega$. Therefore, we can write:

$$\frac{E_{a,FL}}{E_{a,NL}} = \frac{\varphi_{FL} \omega_{FL}}{\varphi_{NL} \omega_{NL}} = \frac{\varphi_{FL} n_{FL}}{\varphi_{NL} n_{NL}} \tag{12}$$

In a permanent magnet DC machine, we know that $\varphi = $ Const. Hence:

$$\frac{E_{a,FL}}{E_{a,NL}} = \frac{n_{FL}}{n_{NL}} \tag{13}$$

$$\Rightarrow \frac{240}{300} = \frac{2400}{n_{NL}} \Rightarrow n_{NL} = 3000 \; rpm$$

Choice (4) is the answer.

22.8 Based on the information given in the problem, we have:

$$n_{rated} = 1000 \; rpm \tag{1}$$

$$P_{cu}^{p.u.} = 3\% = 0.03 \; p.u. \tag{2}$$

$$P_{rot}^{p.u.} = \text{Const.} = 2\% = 0.02 \; p.u. \tag{3}$$

As we know, in a series machine and in the per unit (p.u.) system, the ohmic power loss (copper loss) is equal to the total resistance of machine. Hence:

$$R_a^{p.u.} + R_s^{p.u.} = P_{cu}^{p.u.} = 0.03 \; p.u. \tag{4}$$

Moreover, in the per unit (p.u.) system, we have:

$$V_{t,rated}^{p.u.} = 1 \; p.u. \tag{5}$$

$$I_{t,rated}^{p.u.} = 1 \; p.u. \tag{6}$$

Figure 22.4 shows the electrical circuit of a series DC motor. Applying KVL in the loop for the first condition (rated values):

$$E_{a,rated}^{p.u.} = V_t^{p.u.} - \left(R_a^{p.u.} + R_s^{p.u.}\right)I_{a,rated}^{p.u.} = V_t^{p.u.} - \left(R_a^{p.u.} + R_s^{p.u.}\right)I_{a,rated}^{p.u.} = 1 - 0.03 \times 1 = 0.97 \; p.u. \tag{7}$$

An electric machine has the maximum efficiency if its variable power loss is equal to its constant power loss. Therefore, for a series DC machine, we have:

$$P_{rot}^{p.u.} = P_{cu}^{p.u.} \Rightarrow P_{rot}^{p.u.} = \left(R_a^{p.u.} + R_s^{p.u.}\right)\left(I_{a,\eta_{max}}^{p.u.}\right)^2 \Rightarrow I_{a,\eta_{max}}^{p.u.} = \sqrt{\frac{0.02}{0.03}} = 0.8165 \; p.u. \tag{8}$$

Applying KVL in the loop (the second condition):

$$E_{a,\eta_{max}}^{p.u.} = V_t^{p.u.} - \left(R_a^{p.u.} + R_s^{p.u.}\right)I_{a,\eta_{max}}^{p.u.} = 1 - 0.03 \times 0.8165 = 0.9755 \; p.u. \tag{9}$$

As we know, the induced armature voltage of an electric machine can be determined by using $E_a = k_a\varphi\omega$. In a series machine with linear magnetization characteristics, we have $\varphi \propto I_a$. Therefore, we can write:

$$\frac{E_{a,rated}}{E_{a,\eta_{max}}} = \frac{\varphi_{rated}\omega_{rated}}{\varphi_{\eta_{max}}\omega_{\eta_{max}}} = \frac{\varphi_{rated}}{\varphi_{\eta_{max}}}\frac{n_{rated}}{n_{\eta_{max}}} = \frac{I_{a,rated}}{I_{a,\eta_{max}}}\frac{n_{rated}}{n_{\eta_{max}}} \tag{10}$$

$$\Rightarrow \frac{0.97}{0.9755} = \frac{1}{0.8165} \times \frac{1000}{n_{\eta_{max}}} \Rightarrow n_{\eta_{max}} \approx 1232 \ rpm$$

Choice (4) is the answer.

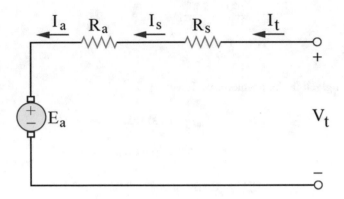

Fig. 22.4 The electrical circuit of solution of problem 22.8

Index

© The Editor(s) (if applicable) and The Author(s), under exclusive license to Springer Nature Switzerland AG 2022
M. Rahmani-Andebili, *DC Electric Machines, Electromechanical Energy Conversion Principles, and Magnetic Circuit Analysis*,
https://doi.org/10.1007/978-3-031-08863-6

Printed in the United States
by Baker & Taylor Publisher Services